SHARED WATERS, SHARED OPPORTUNITIES

Hydropolitics in East Africa

SHARED WATERS, SHARED OPPORTUNITIES

Hydropolitics in East Africa

Bernard Calas & C.A. Mumma Martinon

French Institute for Research in Africa
Jesuit Hakimani Centre
Mkuki na Nyota Publishers Limited

Published by:
Mkuki na Nyota Publishers Ltd.
P. O. Box 4246
Dar es Salam, Tanzania
www.mkukinanyota.com

French Institute for Research in Africa (IFRA)
P.O.Box 58480 - 00200,
Nairobi, Kenya.
www.ifra-nairobi.net

Jesuit Hakimani Centre (JHC)
P.O.Box 214 - 00625,
Nairobi, Kenya.
www.jesuithakimani.org

© IFRA, 2010

Maps: Amélie Desgroppes

ISBN 978-9987-08-092-2

Contents

SECTION ONE

Conflicts and Management

SECTION TWO

Management and Practices

With deep respect,
For Prof. Jean-Pierre RAISON of Nanterre University, France.

Acknowledgements

Profound appreciation to the following persons who greatly assisted in the realisation of this publication.

For editorial work:

Zacharia Chiliswa
Ndanu Mung'ala
Delphine Lebrun
Julie Houpon

For design and Layout:

Zacharia Chiliswa

For maps:

Amélie Desgroppes

And to all IFRA staff for the valuable support they offered, in their various capacities, during this project.

(Centre National de la Recherche Scientifique),
Paris, France

- Prof. Marcel Rutten, marcel.rutten@ consunet.nl / rutten@ascleiden.nl
Geographer & Head of the Economy, Environment & Exploitation Research Group of the African Studies Centre, Leiden, The Netherlands

- Dr. Stéphanie Duvail, stephanie.duvail@ ird.fr
Geographer, Institute of Research for Development (IRD), UMR 208, IFRA, Nairobi, Kenya

- Dr. Olivier Hamerlynck, olivier. hamerlynck@wanadoo.fr
Ecologist, Centre for Ecology and Hydrology, Wallingford, Crowmarsh Gifford, Oxfordshire, UK

- Dr. Heather Hoag, hjhoag@ufsa.edu,
Historian, University of San Francisco, USA

- Prof. Pius Yanda, yanda@ira.udsm.ac.tz
Geographer, Institute of Resource Assessment, University of Dar es Salaam, Tanzania

- Dr. Jean-Luc Paul, jlpaul@univ-ag.fr
Anthropologist, Institute of Research for Development (IRD), UMR 208, IFRA, Nairobi, Kenya

- Delphine Lebrun, delphinedglebrun@ gmail.com
Geographer, Nairobi, Kenya.

- Dr. Judith Nyunja, JNyunja@kws.go.ke
Senior Scientist, Kenya Wildlife Services (KWS), Nairobi, Kenya

List of Contributors

- Prof Bernard Calas, fracasses@wanadoo.fr
  Geographer & Professor,
  University of Michel de Montaigne -
  Bordeaux 3,
  UMR ADES-DyMSET, France

- Dr. C.A. Mumma Martinon,
  connie_martinon@yahoo.co.uk,
  Conflict Prevention Analyst,
  International Peace Support Training
  Centre (IPSTC)Westwood Park Karen -
  UNDP, Kenya

- Abdullahi Elmi Mohamed ,
  abdullaahielmi@gmail.com,
  Somali Centre for Water and Environment
  (SCWE),
  Mogadishu University (MU), Mogadishu,
  Somalia

- Hussien M. Iman, husseinimaan@yahoo.
  com,
  Somali Centre for Water and Environment
  (SCWE),
  Mogadishu University (MU), Mogadishu,
  Somalia

- Kakeeto Augustine Richard,
  gustekakeeto@yahoo.com.

- Hope T. Chichaya, htchichaya@yahoo.
  co.uk.

- Andrea Nicodemo, soliana2002@yahoo.
  com.

- Catherine N. Ndungu, catherine_ngigi@
  yahoo.com.au.

- Daniel Peter Lesooni, peterdanleso@
  yahoo.com.

- Ochwoto Christopher Ogachi,
  chriskenosc@yahoo.com.

- Matthias Tagseth, mattias.tagseth@svt.
  ntnu.no,
  Department of Geography,
  Norwegian University of Science and
  Technology,
  N-7491- Trondheim, Norway

- Haakon Lein, haakon.lein@svt.ntnu.no,
  Department of Geography,
  Norwegian University of Science and
  Technology,
  N-7491- Trondheim, Norway

- Silas Mutia M'Nyiri, silasmutia@yahoo.
  co.uk
  Chief chemist, Ministry of Water and
  Irrigation,
  Nairobi, Kenya

- Samuel O. Owuor, samowuor@yahoo.
  com,

- *Senior Lecturer Department of Geography*
  & Environmental Studies,
  University of Nairobi, Kenya

- Dr. Joseph Onjala, jonjala@uonbi.ac.ke /
  onjalajosef@yahoo.com,
  Research Fellow at the Institute for
  Development Studies (IDS), University of
  Nairobi, Kenya

- Dr. Mathieu Mérino, mateu.merino@
  voila.fr,
  Researcher associated with the Center for
  Research and Studies in Africa (CREPAO),
  University of Pau, France

- Dr. Alphonce G. Kyessi, kyessi@aru.ac.tz,
  Research Fellow in the Institute of Human
  Settlement Studies, College of Lands and
  Architectural Studies, University of Dar es
  Salaam, Tanzania

- Dr. Marianne Kjellén, marianne.kjellen@
  humangeo.su.se
  Research Fellow at the Stockholm
  Environment Institute,
  Sweden

- Dr. Jean Huchon, jeanohuchon@yahoo.fr
  Researcher, Institute of Research for
  Development (IRD),
  Marseille, France

- Dr. Janick Maisonhaute, janick.
  maisonhaute@free.fr
  Researcher CNRS

Introduction

Water and Life

C.A. Mumma Martinon, PhD

The importance of watercourses to human life and development cannot be overemphasised. Throughout human development, watercourses have played a major role as the medium of communication, trade, agriculture, fishing, recreation, tourism, culture, and location of human settlements. The importance of watercourses has become even more significant in the light of the ever-increasing human population. As the human population increases, a corresponding need arises for fresh water for domestic consumption and for agricultural and industrial uses.[1]

Most of the world's largest rivers are international and with the formation of the confederation of independent states, the number is growing. Nearly 40% of the world's population is dependent on about 43 international rivers in the Americas, 20 in Europe, 27 in Africa and 50 in Asia.[2] 47% of the world (excluding Antarctica) falls within shared rivers and lake basins, from a high of nearly 60% of the areas in Africa to South America to a low of about 40% in North and Central America.[3] For instance, the Amazon Basin is shared by seven nations, the Danube by more than eight, both the Niger and the Nile by more than seven, the Rhine by seven, the Zaire by nine and the Zambezi by six nations. India and Bangladesh haggle over the Ganges-Brahmaputra, while Mexico and the US do the same over the Rio Grande; Egypt, Ethiopia and the Sudan over the Nile; Iraq, Syria and Turkey over the Euphrates and the Tigris.

The situation above shows that any management of water should be done in a co-operative manner since water does not respect territorial boundaries and the consequences of its use or removal by upstream countries are immediately felt downstream.

Because of population increase and increasing demand of water for agriculture, industry and urbanisation, a large number of countries fall into the category of 'water scarce'[4] nations. Observers say that by the year 2025, 48 countries in the world will be severely short of water and that the people on earth will not have access to clean water supplies. For instance, in Africa alone, 300 million people - a third of the continent's population - have already started living under water scarcity situations as of the beginning of the millennium.[5] Unless something is done to thwart the problem, twelve more African countries will join the thirteen that already suffer from 'water stress' or 'water scarcity' about a quarter century from now, Since the amount of available freshwater is, however, not increasing, there is now more than ever before, an urgent need to take effective measures for proper management of freshwater resources, including their protection and preservation from activities that cause their pollution.

It is within this background that in Nairobi, Hekima College in collaboration with the Jesuit Hakimani Centre and French Institute for Research in Africa (IFRA) hosted the Hekima College Water Day Academic Seminar on 20 March, 2009 with the theme 'Shared Waters, Shared Opportunities'. Here, critical issues affecting shared water resources in Africa and different perspectives from other parts of the world were discussed.

This book is therefore, a result of the research and presentations from renowned scholars, researchers and experts from different countries and the students of the Institute of Peace Studies and International Relations – Hekima College. The book examines some of the conflicts surrounding water in Africa and loopholes in the existing institutional frameworks. It highlights the existing management mechanisms locally and illuminates the different practices for effective water management towards the reduction of political and natural resource tension in the region.

The main significance of this book is the time period for the water conflict management issue. For the most part of 2009, Kenya faced severe water scarcity. The water crisis in Kenya led to perpetual water and electricity rationing, a rise in food prices, drought, the death of both humans and animals, and a proliferation of water borne diseases among other problems. In a city like Nairobi, many taps went dry and many households went for several months without running water, resulting in exorbitant water prices, which in most cases the majority could not afford.

In 2010, various parts of the country were faced with flooding, which saw many people and cattle dying, homes and roads being swept away leaving many homeless and displaced. This, coupled with huge landslides in affected areas, left many wondering which is the best way to manage water and thus water related conflicts, not only in Kenya but in other parts of the Eastern Africa region, as it has become clear that both scarcity of water and excessive water can be disastrous if not managed properly.

Thus, the issues being discussed in this book are timely and may highlight some of the critical issues which have been underlying the major water crises in Kenya and the Eastern Africa region at large. More information on these issues could be a major step in providing solutions to some of the water conflicts in Africa and the world over in general.

Again, many authors have spoken on international water conflicts.[6] Recent studies, particularly in the field of Environmental Security, have focused on the potential conflicts of these international waters.[7] In addition to this, there has been a raging debate over the possibility of war over shared waters and researchers have attempted to collect data and analyse the issue on a regional and global scale.8 The discussions presented in this book open a debate into the discussions of water issues in Kenya and in the region at large which could be a major step towards joining the global debate over water.

Notes

1. Idris, D., Sinjela, M. (1995). The Law of Non-Navigated Uses of International Watercourses. *The International Law Commission's Draft Article: An overview.* Vol. 3. 1995, p. 84.

2. Kliot, N. (2001). Development of Institutional Frameworks for the Management of Transboundary Water Resources. *International Journal, Global Environmental Issue.* Vol. I, No. 3 & 4, pp. 306-326.

3. Biswas, A. K. (1992). Water for the Third World Development. *Water Resources Development,* Vol. 8, pp. 34 - 36.

4. Scientists define 'water scarcity' as the availability of less than 1000m³ of water per person per year, while 'water stress' as the availability of less than 500m³ of water per person per year (BBC News Online, 15 November, 1999).

5. Tafesse, T. (2000). *The Hydropolitics Perspective of the Nile Question.* Available at http://chora. virtualave.net/tafesse-nile.html (28 May 2009).

6. Homer-Dixon, 1994; Westing, 1986; Kliot, 1994 and Gleick, 1993.

7. Westing, 1986; Gleick, 1993 and Homer-Dixon, 1994.

8. Meridith et al. (2003) p.164; Gleick (1993) p. 79.

The Politicisation of Water in East Africa[1]

Bernard Calas

This work is the result of a long, certainly too long a delivery, since its foundations were laid in 2004 by Professor Bernard Charlery de la Masselière, who was then director of IFRA. A number of articles were proposed at the time that have since been completed in two series of contributions : on the one hand, those on trans-boundary basins presented during celebrations to mark the World Water Day on 20 March, 2009 organised by Dr. Connie Mumma Martinon of Hekima College, Nairobi and funded by IFRA, and on the other hand those written by Professor Marcel Rutten of the Leyden University presented during a seminar to mark the World Environment Day on 6 June, 2008 organised by IFRA in partnership with the University's Geography Department and the Alliance Française of Nairobi with funding from the Alembert Fund of the French Ministry of Foreign Affairs.

Though we risked imperfection, the quality of the papers convinced us to present all of them to the reader. These articles were completely reviewed and updated by their authors in the fall of 2009 (September - November 2009). We believe that even though the project is old, presenting it to the reader in its entirety was justified by the assortment of sources and the fact that since 2004, the issue of water has emerged with great vigour. Emphasis was placed on issues of water, hydraulics and hydrology in Eastern Africa with the intention of providing an overview of the variety of hydro-political situations.

Why does this work focus on East Africa? Demarcating East Africa from a hydrological point of view is evidently subject to the use of a purely academic style. Indeed, even though the 16 papers in this collection take an interest in situations that are very localised to East Africa, this is more as a result of a compilation informed by contiguity in space rather than any systemic or intellectual coherence. Nevertheless, beyond this undisputed preliminary observation, the elements of a factual reply to the question: "Is there an East African hydrological unit?" do not fail to elicit interest since they make it possible to specify the hydrological nature of the zone covered by this collection.

Average Annual Rainfall in East Africa

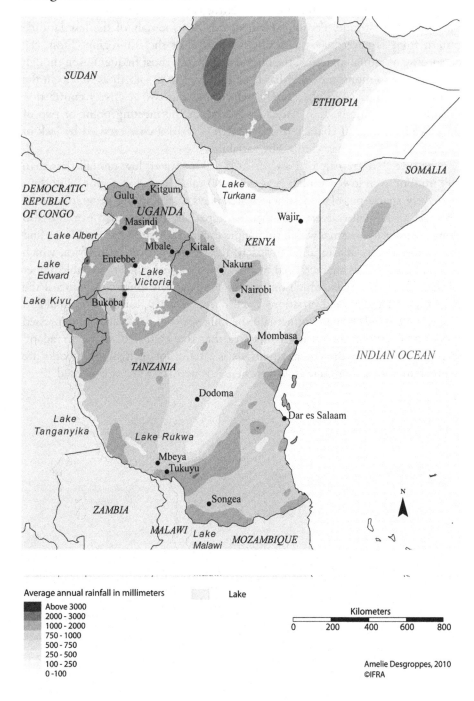

Average annual rainfall in millimeters

- Above 3000
- 2000 - 3000
- 1000 - 2000
- 750 - 1000
- 500 - 750
- 250 - 500
- 100 - 250
- 0 - 100

Lake

Kilometers

| 0 | 200 | 400 | 600 | 800 |

Amelie Desgroppes, 2010
©IFRA

On a continental or even universal scale, the rainfall map customises the East African space, which is defined by its relative lack of rainfall that is unique to this equatorial latitude. Poorly explained, this dry anomaly of the low latitude eastern front is nevertheless a distinctive feature of the sub-region. From the perspective of rainfall patterns, the bimodal pattern is most frequent even though the tendency to mono-modality has been observed in the north as well as in the south. On a larger scale, the rainfall map brings out two relatively contrasting hydrological situations. East Africa is, indeed, at the meeting point of two of the three hydrological characteristics of Africa:[2] "Africa characterised by lack of rainfall" and "Africa characterised by variability".

The first characteristic of East Africa is lack of water, low rainfall (less than 700 mm) or even lower rainfall (less than 400 mm), found in regions with low, often very sparse population, where water plays a structuring geographic role. This lack of rainfall converges in the east, especially in the north-east and also along a semi-arid diagonal which sideswipes the north-eastern parts of Kenya and the southern parts of Tanzania. Inversely, the second part of East Africa, which is more populated, is characterised by not so exceptional but very variable higher rainfall. "Raw materials, which on average are relatively abundant, conceal the huge seasonal and annual variations".[3] Thus, like in 2010, torrential rainfall and catastrophic floods brutally punctuate the long spells of dry years and depressed flows (2006-2009). "It is within this area that the 'ability' of States to 'adapt' to highly variable hydrological situations is vital: a lot of hydraulic works are required to ensure that there is water where it is needed, when it is needed".[4]

Figure showing Climographs of East African Meteorological Stations

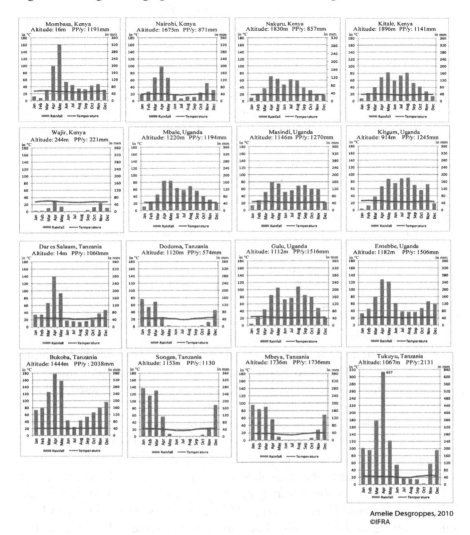

Amelie Desgroppes, 2010
©IFRA

East Africa covers these two types of space and its borders overlap with the watersheds of the three main African rivers: the Nile to the north-west, the Congo to the west and, to a lesser extent, the Zambezi to the south. Kisangani, the continental terminal of the routes fed by the Indian Ocean, is located on River Congo; similarly, Juba, the northern terminal of the routes originating from Mombasa, is situated quite close to the banks of River Nile. Bujumbura, an East African capital, is situated on the shores of Lake Tanganyika which pours into River Congo. Tukuyu, on the southern extreme of Tanzania, faces Lake Malawi which pours into River Zambezi through River Shire. The region can therefore not be accorded the coherence of a watershed – though this is quite debatable.

7

However, the rugged relief – caused by the ridges of the Rift Valley, a brittle tectonic that alternates between valleys and mountainous highlands – and very frequent effusive volcanic activity, the successive reorganisation occasioned by the flows during geological times and the induced compartmentalisation complicate the flow of water and make it possible, in a way, to distinguish an East African hydraulic space from that of its neighbours. Indeed, the tectonics and volcanic activity fragment the watersheds – on a limited whole (especially in the east) – cause an original endorheism (especially in the central region) and a considerable lacustral accumulation (especially to the west of the region specifically called the Great Lakes of Africa: Lake Victoria is a 2,910 billion m³ reservoir!). These three factors combined give an outline of a – dotted and very disputable – border between the hydrological East Africa and the rest of the continent.

From the shores of the Indian Ocean, it follows the northern border of the Juba-Shebeli watershed to the north-east, the shores of Lake Turkana and Bar el Djebel situated in upper Juba (originating from the confluence of the Albert Nile and Aswa) to the north-west, the border of the Rift Valley great lakes to the west (Kalemie, like Juba, is situated on the border) and the border of the Ruvuma watershed to the south. This border only has weak consistency insofar as it does not make the region unique with very clear characteristics and imposing in itself. It is therefore rather low-profile in between other regions where the hydrological characteristics and/or the sensitivity of the hydro-political issue is much clearer. Gérard Prunier shows how the great political game which is intertwined with the issue of the Nile first between Egypt and Ethiopia, with Sudan and Somalia as secondary partners, justifies this partitioning into two of the Nile watershed and the inclusion of its upper part in the Great Lakes of Africa.[5]

Our study area therefore juxtaposes two hydrological characteristics of Africa: to the west, the Great Lakes Africa and the watersheds of the largest rivers, the Nile and the Congo, and to the east, small coastal rivers and endorheic flows. This partitioning according to the types of flow corresponds to partitioning according to rainfall patterns: to the east and especially to the north-east, semi-Sahelian Africa that lacks rainfall, as opposed to Africa characterised by variability, which is unique owing to the variability that is mitigated by oreography as well as the convergence of the easterlies and the western Congolese flows, to the west. The intermediary situation at the intersection of the two is situated east of the Kenyan highlands i.e. Central Province – Kikuyu land, Embu and Meru – which overlook the arid lower plains. Indeed, one of the unique characteristics of East Africa is found in the speed of rainfall gradient due to the influence of topographical gradients on the compartmentalisation of relief. Shelter situations, exposure to the wind, oreographical isolates and presence of large water walls put into perspective the climate pallet and complexify both the rainfall and the flow map. East Africa does not have the well - organised west-western or southern latitude.

Figure showing Catchment Basins in East Africa

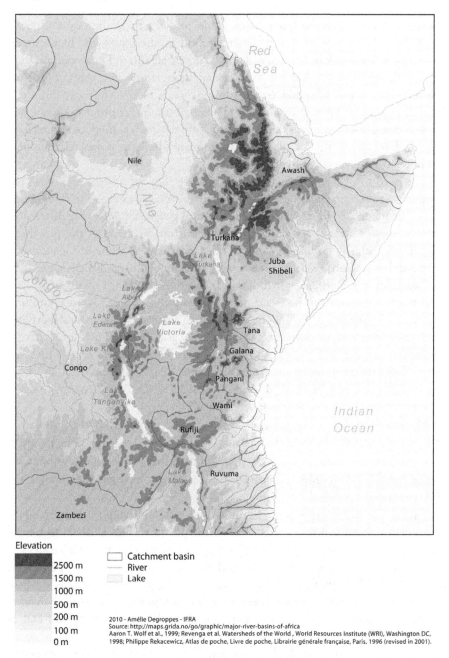

Elevation

2500 m
1500 m
1000 m
500 m
200 m
100 m
0 m

☐ Catchment basin
— River
▨ Lake

2010 - Amélie Degroppes - IFRA
Source: http://maps.grida.no/go/graphic/major-river-basins-of-africa
Aaron T. Wolf et al., 1999; Revenga et al. Watersheds of the World , World Resources Institute (WRI), Washington DC,
1998; Philippe Rekacewicz, Atlas de poche, Livre de poche, Librairie générale française, Paris, 1996 (revised in 2001).

Kenya, Uganda, Rwanda, Burundi, Tanzania, as well as southern Somalia and Ethiopia, are included in the boundaries of the study, therefore showing the State apparatus responsible for managing resources. Indeed, the issue of water is as much if not more of a policy issue as it is a physical issue. However, still according to David Blanchon, of these States, only Kenya appears as a "secondary hydro-hegemonic power"[6] while the other states feature as marginalised actors. Among the necessary attributes to claiming to be a hydro-hegemonic power are: technical expertise, financial capability, and political support. These states, including Kenya, lack at least one element, if not all three. These weaknesses – human, financial and diplomatic – in State apparatus imply that most of the time it is non-state actors who take over settling of issues of sharing water resources. Even though more often than not and the contributions are proof to this, it is at the watershed level or even at the watershed segment level, at the project area level or at the level area settled by an ethnic community where the power relations play out in relation to access to hydro resources.

The need to plan for an East African hydro-policy on a larger scale than the national scale compels us to accommodate the classification put forward by David Blanchon.[7] Indeed, even though "physical geography infers big differences between these states" it also infers the same within the states at the provincial level. Thus, if the hydraulic dependence ratio, i.e. the proportion of the resources from another [region], is considered, three cases can be distinguished: some areas are "water towers" and enjoy hydrological "location advantage". This is the case with Kenya's Central Province and the Mau Forest, Tanzania's Kilimanjaro and Southern Highlands, as well as the Great Lakes watersheds, namely Albert and Victoria. In contrast, the lower regions, especially arid and semi-arid areas that are ill-equipped to act, like the areas through which Tana, Rufiji, Galana, Turkwell rivers flow, are in a more unfavourable, dependent hydro-political position. Some regions are in a "neutral" position in which, in the absence of large border watersheds, geo-political problems do not arise.

Hydrological risk, in the strict sense of the word, is defined by a combination of a climatic hazard (floods, drought) and vulnerability of a people. The hazard is therefore linked to climatic factors, vulnerability to socio-economic development; drought and flooding only attain catastrophic proportions where the population's ability to adapt is limited. Hydrolog.orgotten.

Finally, the recent crises brought on by high costs of living from 2008-2010, combined with oil and food crises increase the pressure of external actors on the resource. There are many "green energy", agro-fuel projects on irrigated sugar cane farms (Tanzania's Bagamoyo region), and vegetable farm projects by petroleum States (Tana delta region). The district migration figures show the sectors where there has been the greatest water pressure. Even though the growth of urban areas is expectedly above the national average (for example, the population of Dar es Salaam grew more than 20 times between 1950 and 2000!), some arid and semi-arid regions bordering the highly populated areas (northern Kenya especially) as well as high - altitude forests (Mau Forest) whose rates of population growth even more unexpectedly go beyond the national average lead to doubts over the durability of the correlation between water availability and population.

Like in the rest of Africa, the issue of water supply to cities in East Africa

(Nairobi and Dar es Salaam – each with a population of 4 million residents or 10% of the total national population – account for 60% of Kenyan and Tanzanian GDP respectively) is a key factor. Despite the decreasing growth rate of large cities compared to that of medium and small towns, the masses make the number of neo-urban dwellers to be connected and supplied considerable. The thirst of cities is not quenched more so because of poor piping and management of the existing resources than for lack of water. One of the reasons why the big cities do not lack too much water is that they are located in areas with fertile agricultural land, are well watered and their political clout has ensured that policy makers make some minimum investment, which the private sector secured to respond to the demand. Thus, in East Africa, the largest water distribution is aimed at serving the large cities: Nairobi is served by Aberdare streams and wells, Mombasa by the Mzima Springs in Tsavo West and Dar es Salaam by Ruvu River. Kampala, Bujumbura, Kisangani, and Juba pump their water from lakes and rivers. However, due to the low expenditure on infrastructure per capita, Dar es Salaam and Nairobi were in the early 1990s among the worst students of expenditure on infrastructure per capita.

In Nairobi, while 30% of the population is connected to tap water, more than half of the population depends on small-scale water vendours who have organised themselves into cartels at the mercy of gangs which contribute to raising the cost of water. The remaining 20% rely on water tankers. In 2009, following several years of drought, the Nairobi water supply dams, especially Sasuma, were virtually emptied, occasioning drastic rationing of both electricity and domestic water to consumers. There are also problems related to lack of water purification infrastructure. This situation has compelled urban populations to constantly make local arrangements and improvise in order to access drinking water. It is these arrangements and improvisation that were of interest to us as addressed by A. Kyessi and M. Kjellén as well as M. Mérino's contribution on the issue of pollution of urban rivers and the current struggle to combat it.

The hydroelectric production, which is also focused on cities, was one of the excuses for State investment and intervention on rivers. Often associated with it are agricultural irrigation schemes but on a small-scale. Indeed, conclusions made by Raison on the irrigation schemes in French-speaking sub-Saharan countries apply perfectly to the geography of rare large-scale planning in East Africa. Even though the irrigation schemes mainly aimed to provide water for agriculture, it was limited to a few schemes, islands of technocratic despotism and rural centres of growth for example on Rusisi in Burundi, mid-stream Tana River in Kenya, on Pangani in Tanzania. "Territorial control was exercised at the schemes which were seen as restricted areas,"[10] removed from their context. The issue that arises is the relationship between the governor and the governed. The intention of the former is to impose land and technical standards and exercise power, reducing the beneficiary to an ambiguous status. In this regard, contributions on the Rufiji and Tana are very informative. The orographic or physical fragmentation add to territorial fragmentation; none of the actors involved in water management has enough clout and power to claim to regulate the use of water on the entire watershed or along a river. The lack of management and territorial power by each of the actors can be seen in the fact that most of the management schemes were limited in time and in space and more often than not are the size of a small scheme or project. "This tends to introduce axial arrangements along the rivers, which are most of the time prone to conflicts".[11]

The succession of those who have upstream rights downstream tends to make the flow of water the criteria of evaluation of good and bad management. Traditionally, the actors' ability to adapt is evaluated on the basis of their adeptness at playing with the flow, i.e. having supply in mind. It is about reducing flooding and minimum flow (particularly using holding dams), sometimes accelerating them (artificial flooding) and most often while varying the flow and distributing the stocks in space (irrigated crops) or redirecting it towards activities that have been presented or considered strategic by policy makers (energy, industrial production) and the urban consumption pools. The issue that arises therefore is sharing of the resource, supply management and, increasingly, regulation of consumption.

The increase in irrigation is another factor affecting the growing demand for water. That is why a considerable number of papers took an interest in this in various physical and political environments: J. Huchon and J. Maisonhaute in Pokot; O. Hamerlynck, S. Duvail, H. Hoag, P. Yanda and J.L. Paul in Rufiji while M. Tagseth took an interest in the Kilimanjaro slopes. The last contribution on flexibility by D. Lebrun, O. Hamerlynck, S. Duvail and J. Nyunja is particularly important and can be considered a conclusion focusing on the interrelationship of conflicts both at the community and interstate level.

Though the presence of immense water reservoirs has delayed the emergence of the hydro-political issue in highland Africa, this is no longer the case today.[12] The Nile issue has caught up with the upper riparian States. From Gérard Prunier's analysis, it is evident that the upper riparian states (14% of the river's watershed) remain minor actors in the Nile geopolitics but have become more and more demanding, especially Tanzania and Uganda. These States have been targeted, in a charm offensive by Egypt, to incentivise representatives of these States to support Cairo to delay as much as possible the renegotiation of 1959 agreement. However, in February 2004, Tanzania initiated an irrigation project near Mwanza which draws its water from Lake Victoria. The rural irrigation project, whose construction was awarded to a Chinese firm and is intended to help a million small-scale farmers, compelled Egypt – a disputed hydro-hegemonic power – to accept against its will to renegotiate the 1959 treaty within the Nile River Basin Cooperative Framework. By late April 2010, all the upper riparian States, led by Ethiopia, pushed Egypt and Sudan against the wall by signing a new water sharing agreement of the great river, even though the Nile Basin Commission had not yet been established.[13]

The issue of the use of Omo River was resolved exclusively by the Ethiopian and Kenyan elites at the expense of riparian pastoral populations. The Gibe III project – the second largest dam in Africa after the Aswan dam – was initiated and is under construction by an Italian firm following the signing of an over US $2 billion contract with the aim of producing 1,800 megawatts of electricity needed for urbanisation and industrialisation in Ethiopia. However, the Gibe III dam will not be immune to traditional problems facing such large projects in the tropics. Indeed, the resultant water retention for a distance of 150 km will be subject to intense evaporation, the dam will soon be filled with silt and the streaming of flooding of Omo River should lead to salinisation of the river delta and reduce the fish resources in Lake Turkana. Doubts on the viability and sustainability of such an undertaking expressed by environmental NGOs like International Rivers, were dismissed by the head of the Ethiopian Environmental Protection Authority: "The advantages for the country as a whole, the local communities and even our neighbour Kenya far outweigh the little problems that could be caused in the short term". Some 500,000 or so people – mainly nomadic and semi-nomadic pastoralists - (surviving) living downstream and on the shores of Lake Turkana will definitely appreciate the cynicism of such a statement.[14]

But the lessons of the past have been partially learned. That is why the World Bank requires more and more often as a prerequisite for its financing, that these facilities be built with sufficiently low valves to allow their managers to make water releases large enough to trigger artificial flooding necessary to sustain downstream watershed ecosystems. One example concerns the designers of the future Grand Falls Dam,[15] on the Tana hydroelectric facilities in Kenya, to reconcile the interests of different actors: urban, sedentary farmers in irrigated areas, nomads' use of flood lands, fishermen, game wardens, etc. The idea is to hold sufficient water to maintain performance of turbines and produce electricity regularly while opening the possibility of provoking a seasonal artificial flood likely to "mimic" the natural flooding to allow for normal functioning of ecosystems and downstream communities. This did not seem to be the concern of the initiators of Gibe III.

Nevertheless, everyone agrees that it is now time to shift supply management to demand management. "Emphasis on water supply, coupled with weak enforcement of regulations, has limited the effectiveness of water resource management, particularly in developing countries. Policy makers have now shifted from entirely supply solutions to demand management, highlighting the importance of using a combination of measures to ensure adequate supplies of water for different sectors".[16] Demand management aimed at reducing over-consumption, will ensure distribution at the expense of the largest and the well-off consumers for the benefit of the majority and the poorest, two categories of people that do not necessarily overlap. Hydrological taxation and wastage quotas therefore tend to slowly fall into place, subtly but inevitably. These transfers are not only intended to reduce over-consumption but also to reverse the scandal of charging the poor more, for water, than the rich.[17]

In summary, the contributions in this collection are focused on the manner in which East African societies share access to water and try to regulate the tensions that arise from the unequal sharing of this resource. The interest is, on the one hand, in the conflict resolution procedures implemented by supra-state and state actors and, on the other, in the supply and demand management practices associated with two uses of water: urban water for industrial and urban domestic use and irrigation water. This book therefore examines the issue of water with a geo-political approach at two levels: national and local. The ambition is not evidently to come up with an exhaustive hydro-political panorama of the region – it would have necessitated an interest for example in the Mau Forest, Lake Naivasha, to the northern slopes of Laikipia – the plateau where the needs of Meru farmers, horticulturalists, ranchers, and Samburu pastoralists intertwine – if we confined ourselves to Kenya alone. The ambition is to provide a glimpse of the variety of political situations and arrangements "between conflicts and compromise"[18]. In this sense, our book is focused on water, geopolitically crucial, since by 2050 one quarter of humanity will live in a country affected by the chronic shortage of water, making it more necessary than ever to hold this discussion on finding peaceful and lasting solutions to sharing this resource, reducing inequality and losses in transfers and managing demand so as to avoid situations of conflict.

Notes

1.   This introduction draws from 4 main sources: Blanchon D. « La question de l'eau en Afrique: de la variabilité climatique aux tensions hydropolitiques ». *Diplomatie*, 2010; Blanchon D. *L'espace Hydraulique Sud-africain : Le partage des eaux*. Karthala, 2009, 294 p.; Raison J.-P. et Magrin G. (dirs.). *Des Fleuves entre Conflits et Compromis: Essais d'hydropolitique Africaine*. Karthala, 2009, 296 p. and finally Ayeb H. « La Question Hydraulique en Egypte: Pauvreté, Accès et Gouvernance », in Richard & Alii, *Coordinations Hydrauliques et Justices Sociales*, actes du séminaire Pcsi, Nov. 2004, Montpellier, Cirad. It partly sets out to examine to what extent concepts, clustering and constructive reasoning apply to East Africa.

2.   Blanchon, D. 2010. *La question de l'eau en Afrique: de la variabilité climatique aux tensions hydropolitiques.*

3.   Blanchon, D. 2010. This explains why raw materials per capita and per country only give a glimpse of the problems related to water and should be interpreted with caution.

4.   Blanchon, D. 2010. *La question de l'eau en Afrique: de la variabilité climatique aux tensions hydropolitiques.* .

5.   Prunier, G. 2009.

6.   Blanchon, D. 2010. *La question de l'eau en Afrique: de la variabilité climatique aux tensions hydropolitiques.*

7.   Blanchon, D. 2010. *La question de l'eau en Afrique: de la variabilité climatique aux tensions hydropolitiques.*

8.   Raison, J.-P. & Magrin G. (dirs.) 2009. *Des Fleuves entre Conflits et Compromis: Essais d'hydropolitique Africaine.* Paris : Karthala.

9.   Calas, B. 1999.

10. Raison, J.-P. & Magrin G. (dirs.) 2009. *Des Fleuves entre Conflits et Compromis: Essais d'hydropolitique Africaine*

11. Raison, J.-P. & Magrin G. (dirs.) 2009. *Des Fleuves entre Conflits et Compromis: Essais d'hydropolitique Africaine*

12. By 1936, a dam had been constructed across the River Pangani; the great dam in Jinja was constructed in 1954, with a higher volume compared to that of Boulder on the Colorado River, and with the aim of providing electricity to the whole sub-region to supplement thermal plants. Today the dam produces 180 MW. A new dam is under construction in Bujagali, 4 miles downstream from Jinja. In Kenya, it is the works in upstream Tana – the Seven Forks Project – which produce electricity mainly for towns. Only 16% of the Kenya population has access to electricity! Production of electricity is estimated at 1,200 million KWH. The 2009 drought led to a deterioration in the Kenyan electric mix to the extent that the fall in the level of water in the hydroelectric dams compelled the government to urgently call, at an exorbitant cost, for thermal power and coal plant investors! The result was a decline in the share of blue and green energy in the Kenyan electric mix in favour of fossil energy sources.

13. *East African*, 26 April – 2 May, 2010. 'More trouble over the Nile's waters'.

14. *Daily Nation*, 4 May, 2010, p. 8.

15. Grand Falls dam will be a major dam, 110 m high, with a large storage capacity (volume equivalent to one and a half years of flow) and multipurpose: hydropower, water provision, flood control, large irrigation projects downstream and recreational use upstream. There is also a plan to build two large canals to supply water for the adjacent river basins to the North and to the South (Interbasin water transfer). The impact will obviously be enormous.

16. *GEO 3*, p. 151.

17. IFRA-JKUAT. 2010. *Evaluation Report Assessment of Impacts of Ablution Blocks Project in Informal Settlements of Nairobi*. AFD-AWSB, p. 10 and p. 48. Usually, in Nairobi, the slum dwellers of Kibera, Korogocho and Mukuru, of whom 75% buy their water from kiosks, pay six times more for water than residents of Parklands, Lavington or Runda. 20 litres of water, per jerry can, is worth 2 shillings in the slums as opposed to 30 cents from a tap in the formal areas. Consequently, for example, during the water shortage of 2009, in certain areas of Mukuru, gangs in the slum sold a jerry can of water at 30 or even 40 shillings.

18. Raison, J.-P. & Magrin G. (dirs.) 2009. *Des Fleuves entre Conflits et Compromis: Essais d'hydropolitique Africaine*.

SECTION ONE

Conflicts and Management

Competing Models of Water Resource Management and Their Implications Using the Example of Pangani River Basin in Tanzania

Mattias Tagseth and Haakon Lein

Introduction

From the first part of the 1990s one saw many calls to give attention to problems of water scarcity[1] and an apparently emerging global water crisis[2]. This debate led to the emergence of an 'international consensus' over key principles for sound water - management, based on the so-called Dublin principles[3] developed and interpreted through a series of conferences and policy papers, including the 1992 Earth Summit[4] and various reports and policy papers prepared by international institutions.[5] These documents deal with a large number of water-related issues, but three main principles can be found in these strategy documents. Firstly, the river basin is perceived as the appropriate and 'natural' region for water resource management. Secondly, water should be treated as an economic good. This will, it is argued, lead to more efficient water utilisation under conditions of scarcity. A third key element is that water should be managed at what is termed 'the lowest appropriate level'. These principles for sound water management, and especially the idea that water should be treated as an economic good are controversial and have led to a debate on whether water can be treated as any other commodity or rather should be treated as a basic human right.

The paper presents and discusses different models or approaches to water management, the 'state centred', the 'market-based' and the 'community-based' (Table 1). Each provides different answers to how and by whom, water resources can best, and should be managed. These three approaches are grounded in development ideologies described as 'classic', 'neoliberal' and neo-populist'. They are advocated by different professions and focus on different actors as well as scales of operation.

These models are used as a basis for discussing water policy and water management reform processes in Tanzania, and their implementation in the Pangani River Basin. The main argument presented in the paper is that policy and the activities of the river basin authorities continue to reflect a state-centred, top-down bureaucratic approach to water management, with colonial roots more than they reflect attempts to introduce a functioning water market or to secure local communities' active participation in water management. Conflicts between communities and the water bureaucracy, between a 'modern' management system drawing on global and professional discourses of water management and an 'indigenous' water management system, over what constitutes 'proper' management of water are central to the water management problems in this basin. While the planned implementation of intermediate organisations at the catchment scale may not provide a quick solution, they could become important as forums for the negotiation between forms of water management drawing on different models.

Three Models of Water Management

State Management

A key element in the state management or the 'technocratic' model of water resource management is that water resources should be subject to public planning and management, preferably within the framework of river basin planning. River basin planning, inspired by the Tennessee Valley Authority (TVA) model, has since the 1960s been applied in several African countries. Adams[6] argues that this model initially became so popular both because it appealed to the idea of modernisation, rationality and planning as well as it promised to solve a number of development problems (hydropower, irrigation and general rural development) in an integrated manner.

This technocratic management model is based on the notion that water belongs to the state and that water should be allocated by the state through administrative water rights or licences. Also, due to the nature of the resource, water management should be carried out within the framework of the river basin reaching across other administrative structures. A main purpose of water fees is to recover costs of providing water (infrastructure) and the operation of water management authorities. Furthermore, the model of water management is based on a strong notion of expert decisions and the ideology that water, societies and humans can be managed so as to produce optimal solutions. Planning, management and conflict resolution are tasks to be carried out by water management authorities having overview of both available water resources and clear ideas about the optimal way of allocating them. These organisations may be governed by boards, often involving different 'stakeholders' as representatives of various interests groups or sectors. The dominant professional perspective is hydrological and water management is primarily seen as a task for hydrologists and to some extent irrigation experts.

Although popular the results of river basin projects in Africa have often been quite disappointing and this can be attributed to a number aspects of both design and implementation.[7] Nevertheless, many still seem to argue that the basic idea is good. It is simply the implementation that is the problem. The approach falls within what Blaikie[8] has termed the classical, statist top-down approach to rural development and environmental management. This development approach has been challenged in recent decades by both neoliberal and neo-populist development paradigms.

The Market-based Model

The market-based (or neoliberal) model, presents a fundamental critique of the technocratic model outlined above. This critique is primarily related to basic questions such as who should make decisions and how water allocation should be made.

The most controversial issue in global water reforms discussed in the introduction is the argument laid down in principle 4 in the Dublin Statement on water and sustainable development, where it is stated that 'water has an economic value in all its competing uses and should be recognized as an economic good'. Perry[9] identifies three main reasons of water charging: firstly, to recover the cost of providing the service; secondly, to provide an incentive for the efficient use of scarce water resources; and thirdly, water charges can be used as a benefit tax on those receiving water services to provide potential resources for further investment to the benefit of others in society. The idea of a water fee or tax to cover the cost of providing services (points 1 and 3) may be controversial and difficult to carry out in practice, but can easily (at least in theory) be incorporated into the traditional river basin model. The second point is more complicated, as it requires some kind of functioning water market.

The market model for water management is based on the very fundamental neoliberal argument that while markets may not be perfect, they are certainly better than bureaucrats and politicians in allocating scarce resources for economically efficient outcomes. Water can and should, it is argued, be treated as commodity to be traded at auctions or in other ways. Those able to pay the most, assumingly because they get the highest return on water, should get water; leading to an economically optimal allocation of a valuable and scarce resource. A water market can be organised in different ways and in this model water management will basically be about developing a legal framework and setting up functioning water markets based for instance on tradable water rights.[10] Within this model there is little room for river basin authorities other than for monitoring and overseeing that the market works. Developing a system of tradable water rights may be seen as a way of empowering water users, providing security of water right tenure and providing incentives to consider the full opportunity costs of water.[11] As a market can be seen as a very decentralised system of decision-making, it somehow also fulfils the requirement of another core principle of the Dublin Statement, that water should be managed at the 'lowest appropriate level'.

This approach to treating water as a commodity to be traded on a market has been met with harsh criticism both for providing a reductionist view of what is a truly multifaceted resource and because water has strong symbolic and cultural elements/values. Others have argued against giving private investors control over this 'blue gold' and claim that access to water rather should be seen as basic common good and as a human right. Considering the experience with water pricing, Molle and Berkoff claim that while some cases have been a successful means of cost recovery, '(t)he impact of water charges on efficiency has, in contrast, remained almost entirely elusive'.[12]

The Community-based Model

The statist, technocratic approach to water management has also been challenged by the neo-populist ideas that gained ground in the development discourse from the 1980s, emphasising the need for community participation in development and natural resource management.[13] This led to calls for more creative interactions between 'indigenous' and 'scientific' knowledge systems in natural resource management.[14]

The basic idea here is that water management should be organised at community level. The community, taken for instance as a village, a water users' association, etc., should control the resource and decide on water allocation. The 'members' have the right to utilise the resource, but there is usually no individual ownership and rights to water are embedded in a system of reciprocal rights and obligations. In order to get water one has to fulfil certain obligations (e.g. be a member of the village, contribute in construction or maintenance of infrastructure). Once these obligations are fulfilled, a person has a claim to water along with other members of the community. 'Payment' for water takes place mainly in the form of labour and sometimes other contributions needed for running the system. Various types of community organisations, involving local leaders and village elders, decide principles and rules for water allocation and negotiate in cases of conflicts. Many existing water management systems are in practice community managed. The crucial point from neo-populist development discourse is not so much the insight that groups of irrigators are capable of running their own irrigation schemes, but the vision that these models and capabilities can be successfully mobilised or replicated in outsiders' quest to improve water management. For instance Ostrom's Common Pool Resources theories[15] can be seen as providing a basis for 'crafting' (i.e. designing) new community-based natural resource management institutions. In the water sector this idea is commonly applied in the form of setting up Water Users Associations (WUAs).

There is no doubt that many community-based water management systems may be characterised as quite successful, for instance in sustaining or even improving rural livelihoods or in handling conflicts. However, they may face problems for instance in including non-traditional uses (industry, hydropower, estates), and in dealing with water management issues beyond the local scale.

TABLE 1: THREE MODELS OF WATER MANAGEMENT

	Model		
Issues	**Market**	**State**	**Community**
Principal agent	Market Judiciary	State, Executive Planner, expert	Community Civil society Groups of water users
Ownership of water	Individual property, Private enterprises	State ownership	The community, users
Mechanism for allocating water	Access to water through purchase of a right in a market	Access to water through bureaucratic allocation of licences water rights	Access to water through participation in scheme, inheritance or usufruct
Resource Mobilisation	Water fees and private investments	Taxes/water fees to government/ water management authority	Labour and other contributions to local water users groups
Ways of solving conflicts	Market/ Judicial Courts Highest bidder	Executive: Boards representing 'stakeholders' or government agencies Expert decisions	Civil society: Water committees, general meetings, hearings villageelders
Scale/Regional focus	Individual user	River basin	Village/Community Watershed
Dominant professional perspective	Economists	Hydrologists Engineers	Farmers NGO professionals

Source : Author's compilation.

Tanzanian Water Laws and Policies

The first water control system in what is now Tanzania was established by colonial authorities in the early 20th century. A main purpose was limiting the use of water among the native inhabitants while at the same time securing access to water for European settlers.[16] A draft water ordinance was prepared during the period of German rule, but the first water law was approved under British rule in 1923. A new ordinance was not passed until 1948, and was replaced in 1959 and subsequently again in 1974. The Water Resources Management Act, 2009 was passed, but has not yet entered into force. According to water laws currently in force in Tanzania,[17] water is vested in the state and all water users who want to abstract water from a river or stream must have a 'water right'. Such water rights, or licences, can be obtained from the water officer, who can grant or refuse water rights to any person. If granted, the water right defines the amount of water to be abstracted, for what purpose it can be used, the duration of the right, and also the source of the water. A water right, or part of it, can be transferred along with land only if the water right is made explicitly appurtenant to it. There is no other provision in the current legislation for sale or other types of direct transfers of water rights. A volumetric fee for water use was set in the Water Utilisation Act,[18] in order to provide funding for water management, as first suggested by Teale and Gillman in a study of water management made during construction of hydropower plants in the Pangani River Basin in the 1930s.[19]

Customary or 'traditional' rights to water have coexisted with the statutory water licences since their inception. In communities with a long-standing practice of traditional irrigation, rightful access to water is seen as a matter of usufruct, inheritance or local custom. Tanzania has legal pluralism, and varying customary laws are accepted under the Judicature and Application of Laws Ordinance (No. 453) of 1961 as equal to written law.[20] Despite this, customary rights to water are not recognised in the same way as rights to land.[21] The uneasy coexistence between customary and statutory rights to water has been an issue during legislative review on more than one occasion.[22] Provisions were made for the registration of customary rights in the 1959 Ordinance, but actual registration during the short moratorium was limited and rights accrued by custom or long undisturbed use are not acknowledged by the Water Utilisation Act or the water management authorities. Only administrative water rights already registered under the old water ordinances or the act, are acknowledged as existing rights. As a result, the legislation and the existing records of water rights kept with the Principal and Basin Water Offices are biased in favour of the estate sector, formal irrigation and hydropower over farmer-managed irrigation.[23]

A new Water Policy, replacing the 1991 water policy document, was prepared by the Ministry of Water and eventually sanctioned in 2002.[24] The new national water policy emphasises the river basin as the administrative unit and the vision of integrated water resources management. The previous priorities[25] between alternative sectors have been abandoned in the new policy, where human needs are defined as a first priority before environmental flow, while water allocation for 'other uses will be subject to social and economic criteria, which will be reviewed from time to time'.[26] Furthermore, the policy states that 'in order to realize the objectives of water resources management all water uses, especially water use for economic purposes will be charged for'.[27] The principle that 'water has a value in all its competing uses'[28] and the volumetric pricing of water in order to increase efficiency[29] are also recognised. Trading of water rights is to be introduced as a means of demand management and water conservation.[30] The documents also mention water user and stakeholder participation, which could be interpreted as embracing more neo-populist methods of water management.

Through adoption of the basin as a management unit, through attempting to place a value on the use of water, and through the focus on stakeholder involvement at various levels, the policy document undoubtedly embraces core elements of the 'international consensus' on water management as expressed in the Dublin principles. Integrated water resources management at river basin level, as well as water fees for the recovery of administrative and catchment management costs are, however, in line with old-established suggestions and priorities for this basin.[31] The policy also takes up again the old objective of establishing a universal system of water licences, still not fulfilled after two legislative reviews and fifty years to implement it.[32] The idea that the volumetric water pricing is a method of demand management can be described as novel, yet in line with international neoliberal recommendations on water management. The implementation will take place in the context of other (sectoral) policies and strategies which tend to counteract the centralisation of water management and the use of price mechanisms for allocation. The implementation of the Water Policy will be guided by the National Water Sector Development Strategy,[33] but the process will also be affected by other policies and strategies especially for the irrigation and hydropower sectors.[34]

A legislative review was called for in the water policy paper.[35] The dual system of water rights has been identified as an issue in the legislative process, and the recognition of 'customary rights' was among the objectives of the proposed 'bill for the water resources management act of 2004',[36] whereby customary rights are to be 'of equal status and effect to a granted right'.[37] Within a two-year period, those who have abstracted and used water undisturbed for a period of time are entitled to 'water use permits' upon application to the water officer. After this period has expired, the water officer will be authorised to register them regardless. Various local organisations are to be registered and given a formal status by the officer. An apparent objective is to incorporate customary rights in the registry of statutory licences, thus creating a single path to a legitimate right to water. This is in line with the Water Policy, which states that 'relevant customary law and practice... will be institutionalised into statutes'.[38] This is an improvement over the past assessment that all unlicensed abstractions are illegal, even if it could be seen as a repetition of a strategy that has already failed twice following water legislation reforms in 1959 and 1974. Juma & Maganga[39] have warned that the expectation that customary laws are a transitory system which will die out may not be fulfilled. The neglect of local custom as a legitimate basis for water use may have negative implications for villagers who rely on it, in practice it may cause insecurity of tenure, and in appears to cause problems in implementing Integrated Water Resources Management through strategies that emphasise water licences under statutory law and volumetric fees for water use.[40] The acknowledgement of customary or common law principles of prescriptive rights in the proposal could nevertheless be seen as a movement towards a community-based management model. However, the provisions will also make it easier to incorporate and charge traditional abstractions. Another important suggestion involves authorising the permanent or temporary transfer of water permits, which is one of the prerequisites for establishing a water market. This is clearly a licence to the market-based ideas. Furthermore, provisions are made to set up sub-catchment organisations with boards and water officers. The latter would involve some decentralisation of authority to an intermediate level. These provisions could be interpreted as licences to concerns about participation and representation in water management.

Water Management in the Pangani River Basin

It can be argued that developments in the Pangani River Basin have been important in defining problems and solutions in Tanzanian water management for a long time, due to the intensive utilisation of the available resources for smallholder irrigation, plantations, industrial irrigation, and hydroelectric power. European explorers arriving in the highland areas in the Pangani River Basin from the mid-19th century onwards were impressed by the hill furrow irrigation systems found in the mountain areas in the basin, and by the available water resources.[41] However, from the period of colonisation onwards, the system was often described as wasteful and in need of improvement, reflecting increasing conflicts of interest.[42] The independent state of Tanzania as well as the many donors involved in water development projects in the region over the years appear to have inherited the latter perception.[43] Despite many attempts to control, reform and 'improve' the indigenous water management systems,[44] most irrigation schemes operate under forms of farmer and community management and these continue to play a key role in intensive local farming systems.

Mt. Kilimanjaro, together with other mountain areas, catches orographic precipitation which makes up most of the runoff for Pangani River. The retreating glacier on Mt. Kilimanjaro has become the most visible and powerful icon of climate change from Africa, following a remote sensing study by Hastenrath & Greischar[45] and a much quoted study of palaeoclimate by Thompson et al.,[46] warning that that 'if current climatological conditions persist, the remaining ice fields are likely to disappear between 2015 and 2020'. This is not a new claim, as the process of glacial retreat has been published repeatedly since shortly after Hans Meyer's first climb in 1887. The direct effect of glacial retreat on water availability appears to be negligible, as there is little runoff from the alpine zone with much of the snow and ice evaporating directly to the atmosphere.[47] Nevertheless, with the snows of Kilimanjaro dwindling, the International Union for the Conservation of Nature (IUCN)[48] and others have started to blame the inadequate water supplies on a regional impact of climate change (global warming). A change in precipitation on the mountain has been ruled out by some hydrologists,[49] wheras Mkhandi & Ngana[50] and Hemp[51] have identified negative trends in linear regressions for the period from long term precipitation records from the 20th century. Further studies are needed, but it now appears more likely that changes in climate in the form of a decrease in precipitation during the 20th century and an associated reduction in runoff may be contributing factors to water scarcity in the region.

Policy makers and various consultants have identified poor implementation of existing regulations as a major cause of the impending water crisis in the basin.[52] Among these were the Scandinavian donors, who pushed for changes in water management in the 1990s, due to concerns about the availability of water to run the redeveloped hydropower plant downstream at Pangani Falls.[53] An interpretation of a 'water crisis' was developed, and a set of interventions prescribed and given backing from the Norwegian Agency for Development Cooperation (NORAD) and the World Bank. The Scandinavian involvement in water management reforms in the Pangani Basin lasted through the 1990s, in parallel with World Bank involvement. In more recent years, the basin has become a demonstration site for the Water and Nature Initiative of the IUCN.

The 'water crisis' in Kilimanjaro and the Pangani Basin is multifaceted and experienced in different ways by different actors. Its causes are still open to debate, as are the cures. Thus, in a constructivist interpretation, the emergence and definition of a 'water crisis' in Pangani and Kilimanjaro is linked not only to a possible physical process of desiccation but also to concerns about water for the major development projects in the region, and to the need for state control over water. A conflict perspective on water management at the basin level shows that interpretations of the crisis are associated with the diverging interests of sets of actors. In the Pangani Basin, farmer-managed irrigation competes with irrigation development projects, and irrigation competes with hydropower.[54] Plans for irrigated development in the lowlands have existed since the 1930s.[55] These plans have been reconfirmed on many occasions, but concerns about water for hydropower are among the factors that have slowed the development.[56] Irrigation development in the semi-arid lowland followed at a moderate pace. A Scandinavian preoccupation with hydropower and a Japanese preoccupation with the production of paddy fields led to two major water-intensive development projects within a decade. The first phase of the Japanese-sponsored Lower Moshi Irrigation Scheme opened in 1986 had been funded by 31 million USD by 1992,[57] while the rehabilitation of the hydropower station at Pangani Falls was completed in 1995 at a cost of 850 million NOK (c.136 million USD). This contributed significantly to the water stress in the region, resulting in the establishment of the Pangani Basin Water Office (PBWO) in 1992, as the first river basin management authority in Tanzania. The Pangani Basin's water officer was given the tasks of bringing water use under control through implementation of the water licence in accordance with statutory law, and to implement the new volumetric fees for water use. The PBWO collected data on hydrology and water use, and 1015 abstractions with water rights and 1881 unlicensed abstractions were identified.[58] A crash programme to install several hundred control gates to limit the abstractions of water for farmer-managed irrigation was launched. The unlicensed users were told to apply for statutory water rights and to pay an application fee. Other important objectives were to collect the revised fees for water use in order to fund the activities of the PBWO, and also to reform the traditional organisation of farmer-managed irrigation into registered, standardised 'water users associations', which could be held accountable to the water authorities.

The implementation of river basin management along these lines met with resistance, and what can be described as widespread non-cooperation in the effort to install control structures and in the process of implementing the administrative water licence. The relationship with groups of irrigators became strained during the crash programme, and there were locations which the Basin Water Officer and his staff could not visit safely.[59] There was also resistance and non-cooperation from local government. In the districts with long-established practices of irrigation, the prescribed measures have been met with different notions of the rightful access to water and what constitutes proper management of water. Water itself is understood to be 'a gift from God', owned by no one. The scheme may belong to a family group ('clan'), which appoints the leader of water, while members of the community maintain rights of use. Contribution to the common good (irrigation scheme maintenance), long undisturbed use, and inheritance are principles underlying discussions over rightful access to water. The relevance of the statutory licence in local disputes remains minor, and it is associated with the colonial estates and other outside interests. The local institutions and practices of water management are varied, but in some of the communities with a long history of irrigation, the combination of cash payment and access to water from a canal is seen as immoral, 'a sin against God'. The apparent problem of legitimacy of statutory water rights and volumetric water fees, and the efforts to improve water management through integrated water resources management is still seen as a problem of an inefficient communications strategy[60], rather than as a conflict over resource tenure, procedure and the objectives of water management.

TABLE 2: ELEMENTS OF 'MODERN' AND 'INDIGENOUS' WATER MANAGEMENT SYSTEMS

	'Modern' Water Management System	'Indigenous' Water Management System
Scale	River basin	Local: Irrigation scheme or community
Inspirations, sources, legitimacy	Scientism, developmentalism Inspired by global and professional ideals of rational planning and integrated water resources management	Traditionalism Custom Evolved from local practices
Water rights	Administratively allocated water licences, dependent on payment of regular volumetric fees	Informal rights based on usufruct, inheritance, and needs. Negotiations can allocate proportions of flows dependent on contribution to irrigation scheme
Purpose	Water for hydropower, domestic purposes, industry, agribusiness and smallholder agriculture	Water for peasant crops for subsistence and market, and for domestic purposes
Organisation	Seeks to introduce standardised 'water users' associations'	Varied formal and informal community- based organisations
User obligations	Water fee in cash to River Basin authority	Contribution in kind to the local group of irrigators

Source : Author's compilation.

By 2002–2003, the PBWO was able to collect 75,000 USD, i.e. 38% of the amount due from water rights holders.[61] Due to the cost of monitoring the many water abstractions, a flat rate minimum fee[62] and a practice of billing according to installed capacity rather than actual water use have been adopted. This is rational in terms of cost recovery, as the costs of monitoring and billing these abstractions even on an annual basis in areas with poor road access are considerable, but it undermines the intended role of fees as an economic incentive. The progress in registration of water rights was limited, due to lack of both applications and caution in issuing new water rights. Despite emergency measures in water management, subsequent droughts and competing demands for water for irrigation in the industrial and 'traditional' sectors have led to less flow of water, in recent years, through the turbines than the hydropower plant at Pangani Falls was designed for.[63] The difficulties in establishing a single statutory system of licences may have been underestimated, repeating the failures of the campaigns following the legislative reforms in 1959 and 1974.

Alternative proposals have been put forward by NGOs engaged in assistance to traditional irrigation and the environment,[64] with some support from the

World Bank's Traditional Irrigation Improvement Project and the IUCN. The suggestion is to integrate the varied scheme organisations into catchment organisations controlling a source of water.[65] These could form a basis for a tiered water management organisation and make improved representation of stakeholders on the River Basin Board possible.[66] These principles have been tested only to a limited extent in a few project areas in this river basin,[67] but they will be facilitated by the new water legislation.[68] The water sector development strategy,[69] suggests an organisational structure with financially and administratively autonomous Basin Water Boards funded by user charges (and free to negotiate support independently) and Sub-Catchment Committees with powers delegated from the basin level.

Finally, the objective of re-organising the users into registered formal organisations (water users associations) is reconfirmed. The strategy devises charges for water mainly as a way to overcome inadequate funding for management activities.[70] The Pangani River Basin Management Project supported by the IUCN and the United Nations Development Programme is planning to implement a 'Catchment Forum' for a sub-basin (Kikuletwa) as a pilot project where water users can be given a voice.[71] This forum may not be a quick solution to conflicts over 'good' water management, and if it is insufficiently adapted there is a danger that introduced institutions such as Catchment Water Boards and Water Users' Associations may fail to reproduce the legitimacy of indigenous models and institutions water management.[72] Nevertheless, it could become a place of mediation between plural normative repertoires in water management[73] and between local and river basin scales. Further, it could become an important forum for negotiation between what we describe as a 'modern' model of water management drawing on the state-centred model of water management with some adjustments from global water policy discourse and an 'indigenous' model of water management (Table 2).[74]

Conclusion

The three approaches to water management outlined in this article each provide different answers to how and by whom limited water resources best can and should be managed. The traditional technocratic model found in many river basin water management projects around the world is currently being challenged by both neoliberal and neo-populist ideas.

In the water policy documents[75] as well as in the activities of the PBWO there is evidently some adjustment to more general and global trends in water management reforms, with at least some token references to issues such as water pricing and local participation. The suggestion to legalise the sale of permits is clearly a gesture to the neoliberal model, but with the implementation of the new policy there remains a distinct possibility that 'the value of water' may translate into a water tax and not a competitive market-based pricing ensuring efficient use of available resources in economic terms. The drive towards 'demand management' through the price mechanism has provided arguments for a strategy based on water fees, but in practice this is more about cost recovery for water management rather than about attempting to introduce more controversial and difficult market mechanisms based on ideas of economic efficiency.

Devolution or decentralisation is definitely an issue in African and in Tanzanian resource management,[76] but the impact on the water sector still remains comparatively small. The Dublin principles on public participation and devolution of management to the 'lowest appropriate level' can be, in the Tanzanian policy context, translated into campaigns for the 'creation of awareness' in order to increase compliance with regulations. Consultation, transparency or the transfer of responsibilities to sub-catchment organisations or communities may be more difficult to achieve. The acknowledgement of customary or common law principles of prescriptive rights in the proposal for new water legislation could be seen as a movement away from the technocratic model. The establishment of sub-catchment organisations would involve some decentralisation of authority to a local level, and perhaps some adaptation to local practice in water management related to the neo-populist influence. A water management organisation covering a smaller hydrological region could provide a meeting place between the 'indigenous' and 'modern' water management systems, or a framework for incorporating 'traditional' irrigation into the state management system.

Central to the development of water management in the Pangani River Basin in Tanzania is what can perhaps best be understood as a conflict between a local, community-based water management system and a fairly technocratic water management system. The activities of the Pangani Basin Water Office and the donors supporting it, as well as the new water policy and strategy can probably best be interpreted as a continuation of a traditional top-down bureaucratic approach to water management, rather than representing a renewal in terms of securing local communities' active participation in water management or facilitating the emergence of a functioning market for water.

Notes

1. Falkenmark, Malin & Lundqvist, Jan, & Widstrand, Carl.1990. *Water Scarcity, An Ultimate Constraint in Third World Development: A Reader on a Forgotten Dimension in Dry Climate Tropics and Subtropics.* Tema V Report 14: Linköping: University of Linköping, Department of Water and Environmental Studies.

2. Clarke, Robin. 1993. *Water – The International Crisis.* London: Earthscan. Postel, Sandra. 1992. *The Last Oasis.* New York: Earthscan Publications.

3. International Conference on Water and the Environment. 1992. *The Dublin Statement and the report of the conference, 26-31 January 1992, Dublin.* Geneva: World Meteorological Organization.

4. Johnson, S. P. 1993. *The Earth Summit: The United Nations Conference on Environment and Development* (UNCED). London: Graham & Trotman.

5. World Bank. 1993. *Water Resources Management: A World Bank Policy Paper.* The World Bank. 2003. *Water Resources Sector Strategy: Strategic Directions for World Bank Engagement..* Washington D.C.: The World Bank.

6. Adams, William. M. 1992. *Wasting the Rain: Rivers, People and Planning in Africa.* Minneapolis: University of Minnesota Press.

7. Barrow, Christopher J. 1998. "River Basin Planning and Management. A Critical Review," *World Development, 26,* no.1. pp.171-186.

8. de Haan, Leo & Blaikie, Piers (eds.). 1998. "Paradigms for Environment and Development," in Looking at Maps in the Dark: Directions for Geographical Research in Land Management and Sustainable Development in Rural and Urban Environments in the Third World. Nederlandse Geografische Studies/*Netherlands Geographical Studies, no. 240,* pp. 9–40.

9. Perry, C. J. 2001. *Charging for Irrigation Water: The Issues and Options, with a Case Study from Iran.* International Water Management Institute Research Report 52 Colombo: International Water Management Institute.

10. Rosegrant, Mark W. & Binswanger, Hans P. 1994. " Markets in Tradable Water Rights: Potential for Efficiency Gains in Developing Country Water Resource Allocation" in *World Development, 22,* no. 11, pp. 1613-1625.

11. Schleyer, Renato Gazmuri & Rosegrant, Mark W. 1996. "Chilean Water Policy: The Role of Water Rights, Institutions and Markets" in *Water Resources Development, 12,* no 1, pp. 33–48. Gleick, Peter H. 1998. "The Human Right to Water" in *Water Policy, no. 1,* pp. 487-503. Petrella, Riccardo. 2001. *The Water Manifesto: Arguments for a World Water Contract.* London: Zed Books. Metha, Lyla. 2000. *Water for the Twenty-First Century: Challenges and Misconceptions.* IDS Working Paper 111, Brighton: Institute of Development Studies.

12. Molle, François & Berkoff, Jeremy. 2007. "Water Pricing in Irrigation: The Lifetime of an Idea" in Molle, F. & Berkoff J. (eds.) Irrigation Water Pricing: The Gap between Theory and Practice. *Comprehensive Assessment of Water Management in Agriculture, Series No. 4.* Wallingford: CABI International.p.10.

13. Chambers, Robert. & Pacey, Arnold. & Thrupp, Lori, A. (eds.). 1989. *Farmer First: Farmer Innovation and Agricultural Research.* London: Intermediate Technology Publications.

14. de Walt, Billie R. 1994. "Using Indigenous Knowledge to Improve Agriculture and Natural Resource Management" in *Human Organization, 53,* no 2, pp. 123-131.

15. Ostrom, Elinor. 1990. *Governing the Commons: The Evolution of Institutions for Collective Action.* Cambridge: Cambridge University Press. Ostrom, Elinor. 1992. *Crafting Institutions for Self-Governed Irrigation Systems.* San Francisco: ICS Press.

16. Mwita, D. M. (1975, LLM degree thesis). *Rights in Respect of Water in Tanganyika.* Dar es Salaam: University of Dar es Salaaam.

17. Until the new Water Resources Management Act, 2009 has entered into force, water utilisation is basically regulated by the United Republic of Tanzania Water Utilization (Control and Regulation) Act, 1974 (No. 42), with later amendments providing for water management authorities and fees. Regarding the development of statutory water laws, see Van Koppen, Barbara, Sokile, C. S., Lankford B. A., Hatibu, N., Mahoo, H. & Yanda, P. Z. (eds.). 2007. Water Rights and Water Fees in Rural Tanzania in F. Molle and J. Berkhoff. (eds.). *Irrigation Water Pricing: The Gap Between Theory and Practice.* Wallingford: Cab International. Lein, Haakon & Tagseth, Mattias.2009. "Tanzanian Water Policy Reforms – Between Principles and Practical Applications" in *Water Policy 11, no. 2,* pp. 203–220.

18. United Republic of Tanzania, "Water Utilization (Control and Regulation) Act, 1974".

19. Teale, Edmund O., & Gillman, Clement. 1934. Report on the Investigation of the Proper Control of Water and the Reorganisation of Water Boards in the Northern Province of Tanganyika Territory, November–December 1934. Morogoro/Dar es Salaam: [The Government Printer] 1935.

20. United Republic of Tanzania. 2004. *Review of Customary Water Law Regimes: Issues for the Reform of Water Laws.* Consultancy Services on Reviewing Water Resources Legislation Water Supply and Sewerage and Rural Water Supply & Sanitation by Mawenzi Advocate's Chambers.

Dar es Salaam: River Basin Management (RBM)/ Smallholder Irrigation Improvement Project (SIIP) - Ministry of Livestock and Livestock Development.

21. United Republic of Tanzania. 1994. *Report of the Presidential Commission of Inquiry into Land Matters*. Uppsala: The Ministry of Lands, Housing and Urban Development, Government of the United Republic of Tanzania in cooperation with the Scandinavian Institute of African Studies.

22. Teale & Gillman. 1956. *Report on the Investigation of the Proper Control of Water*. [Government of Tanganyika], Report of the Water Legislation Committee. Dar es Salaam: [The Government Printer] 1956. Scott, Peter. 1962. *The Development of the Pangani River Basin*. London: Sir William Halcrow and Partners. United Republic of Tanzania, *Review of Customary Water Law Regimes*.

23. Tagseth, Mattias, "The *Mfongo* Irrigation Systems on the Slopes of Mt. Kilimanjaro, Tanzania" in Tvedt, Terje & Jakobsson, Eva. (eds.). 2006. *A History of Water. Vol. 1*: Water Control and River Biographies. London: I. B. Tauris. pp. 488-506. Tagseth, Mattias. (2001 thesis). Knowledge and Development in *Mifongo* Irrigation Systems: Three Case Studies from Mt. Kilimanjaro, Tanzania. Department of Geography, Norwegian University of Science and Technology (NTNU). Reed-Erichsen, Morten. (2003 thesis). Negotiations over Rights to Water: The Process of Determining Allocation of Water between Farmers and Estates in the context of Coffee-producing estates in Kilimanjaro Region – Tanzania. Department of Geography, Trondheim, Norwegian University of Science and Technology (NTNU).

24. United Republic of Tanzania, National Water Policy 2002. Dar es Salaam: Ministry of Water and Livestock Development.

25. The former 1991 water sector policy ranked priorities were domestic use, livestock, food production, and industries before hydropower (Informants under the Ministry of Water and the Ministry of Agriculture were interviewed in Dar es Salaam January 1996 and in Moshi, June 2000).

26. United Republic of Tanzania, National Water Policy, 18.

27. United Republic of Tanzania, National Water Policy, 29.

28. United Republic of Tanzania, National Water Policy, 15.

29. United Republic of Tanzania, National Water Policy, 7, 14.

30. United Republic of Tanzania, National Water Policy, 19.

31. Teale & Gillman. 1956. *Report on the Investigation of the Proper Control of Water*. 137, 146. Scott. The Development of the Pangani River Basin.

32. United Republic of Tanzania. 1956. *Report on the Water Legislation Committee*.

33. United Republic of Tanzania. National Water Sector Development Strategy [Circulation draft June 2004] Dar es Salaam: Ministry of Water and Livestock Development, 2004.

34. United Republic of Tanzania, The Energy Policy. Dar es Salaam: Ministry of Water, Energy and Minerals [Government Printer] 1992. The Study on the National Irrigation Master Plan. Dar es Salaam: Japan International Cooperation Agency / Nippon Koei Co. & United Republic of Tanzania, Ministry of Agriculture and Food Security, 2002.

35. United Republic of Tanzania, National Water Policy, 9, 28.

36. United Republic of Tanzania, Proposed Draft Bill for the Water Resources Management Act, 2004. Consultancy Services on Reviewing Water Resources Legislation Water Supply and Sewerage and Rural Water Supply & Sanitation by Mawenzi Advocate's Chambers. Dar es Salaam: RBM / SIIP project, Ministry of Livestock and Livestock Development, 2004.

37. United Republic of Tanzania, Proposed Draft Bill for the Water Resources Management Act, 2004, Paragraph 21(1).

38. United Republic of Tanzania, National Water Policy, 29.

39. Juma, Ibrahim H., & Maganga, Faustin. P. "Current Reforms and their implications for Rural Water Management in Tanzania" presented at the International workshop on African Water Laws: Plural Legislative Frameworks for Water Management in Africa. Johannesburg, 26-28 January 2005.

40. Maganga, Faustin. 2003. "Incorporating Customary Laws in implementation of IWRM: Some Insights from Rufiji River Basin, Tanzania" in *Physics and Chemistry of the Earth, 28,* no. 20-27, pp. 995-1000. Maganga, Faustin P., Kiwasila, Hilda L., Juma, Ibrahim H., Butterworth, John A. 2004. "Implications of Customary Norms and Laws for implementing IWRM: Findings from Pangani and Rufiji Basins, Tanzania" in *Physics and Chemistry of the Earth, 29,* pp.1335-1342. Van Koppen et al. *Water Rights and Water Fees.*

41. Tagseth, Mattias. 2008. "The Expansion of Traditional Irrigation in Kilimanjaro, Tanzania" in *The International Journal of African Historical Studies, 41,* no. 3, pp. 461-490.

42. Griffith, A.W. 1930. Chagga Land Tenure Report. Moshi: Available from the archives of the University of Dar es Salaam, East Africana collection EAF Cory 272. Swynnerton, R. J. M. "Some problems of the Chagga on Kilimanjaro," in *East African Agricultural Journal* (Amani), January 1949, pp.117–132.

43. Daluti, L. R.1994. *Report on the Agro-socio-economic situation in Pangani River Catchment.* Consultancy for NORPLAN. Moshi: Zonal Irrigation Office, United Republic of Tanzania. Water Utilisation and Shortage in the Pangani River Basin. Moshi: Ministry of Water Energy and Minerals,1994. Lein, Haakon. 1998. "Traditional versus Modern Water Management Systems in Pangani River Basin, Tanzania" in de Haan Leo, and Piers Blaikie (eds.), *Looking at Maps in the Dark: Directions for Geographical Research in Land Management and Sustainable Development in Rural and Urban Environments of Third World.* Utrecht/Amsterdam: Royal Dutch Geographical Society pp. 52-64 & Lein, Haakon. 2004. "Managing the Water of Kilimanjaro: Irrigation, Peasants and Hydropower Development" in *Geojournal, 61,* pp. 155-162.

44. Hastenrath, Stefan, & Greischar, Larry. 1997. "Glacier Recession on Kilimanjaro, East Africa, 1912-89" in *Journal of Glaciology, 43,* no. 17, pp. 455-459.

45. Thompson, L. G., Mosley-Thompson, E., Davis, M. E., Henderson, K. A., Brecher, H. H., Zagorodnov, V. S., Mashiotta, T. A., Ping-Nan Lin, Mikhalenko, V. N., Hardy, D. R., & Beer, J. 2000. "Kilimanjaro Ice Core Records: Evidence of Holocene Climate Change in Tropical Africa" in *Science, 298,* no. 5593, pp. 589-593, 580.

46. Lein & Tagseth. 2008. *Tanzanian Water Policy Reforms.* Mölg, Thomas, Cullen, Nicolas J., Hardy, Douglas R., Kaser, G. and Klok, L. 2008. "Mass Balance of a Slope Glacier on Kilimanjaro and its Sensitivity to Climate" in *International Journal of Climatology, 28,* pp. 881-892.

47. International Union for the Conservation of Nature. *Pangani Basin: A Situation Analysis.* Nairobi: International Union for the Conservation of Nature and Natural Resources, Eastern Africa Programme, 2003.

48. Sarmett, Julius D., & Faraji, S. A. The Hydrology of Mount Kilimanjaro: An examination of Dry Season Runoff and Factors leading to its decrease, in Newmark, W. D. (ed.). 1991. *The Conservation of Mount Kilimanjaro.* Gland, IUCN-World Conservation Union, pp.53–70.

49. Mkhandi, S., & Ngana, J. Trend Analysis and Spatial Variability of Annual Rainfall, in Ngana, J. (ed.). 2001. *Water Resources Management in the Pangani River Basin: Challenges and Opportunities.* Dar es Salaam: Dar es Salaam University Press. pp.11-20.

50. Hemp, Andreas. 2005. "Climate Change-driven Forest Fires marginalize the impact of Ice Cap Wasting on Kilimanjaro", in *Global Change Biology, 11,*pp.1013-1023.

51. United Republic of Tanzania, Water Utilisation and Shortage.

52. Rudberg, Tone Spilt vann kan koste Tanzania norsk støtte for 400 mill [Spilt water could cost Tanzania 400 million NOK in support]. NTB-Norwegian News Agency, Oslo, June 29, 1994. Hjorthol, Lars M., Norad-press mot Tanzania [NORAD- pressure against Tanzania]. NTB-Norwegian News Agency, Oslo, May 27, 1994. "Conflicts over Pangani dampen enthusiasm for hydropower in Tanzania", *Development Today.* Nordic outlook on Development Assistance, *Business and the Environment. 5, no. 12-13,* (1994) 1 & 6. Bryceson, Ian, Norske kraftutbyggere i Tanzania [Norwegian constructors of hydropower in Tanzania]. In *Foreningen for internasjonale vann og*

*skogsstudie*r (FIVAS) (Ed.) *Kraft og konflikter. Norske vannkraft-utbyggere i den tredje verden*. Oslo: FIVAS, 1994, 97-104.

53. Mujwahuzi, Mark R. Water Use Conflicts in the Pangani Basin: An Overview, in Ngana, James O. (ed.). 2002. *Water Resources Management in the Pangani River Basin: Challenges and Opportunities*. Dar es Salaam: Dar es Salaam University Press, pp.128-137.

54. Buckland, L. L. R, *Water Executive Annual Report*. Dar es Salaam: [Government Printer]1939.

55. Teale & Gillman, *Report on the Investigation of the Proper Control of Water*.

56. Beez, Jirgal. *Die Ahnen essen keinen Reis. Vom lokalen Umgang mit einem Bewässerungsprojekt am Fuße des Kilimanjaro in Tansania. Bayreuth African Studies* Working Papers No. 2. Bayreuth: Universität Bayreuth Institut für Afrikastudien & Kulturwissenschaftliches, 2005.

57. Pangani Basin Water Office database 1994, Hale: Pangani.

58. Interviews with Ministry of Water personnel at Ubungo, Dar es Salaam, Hale & Moshi 18 August,1995 to 20 February, 1996.

59. United Republic of Tanzania, *National Water Sector Development Strategy*.

60. Regional Hydrologist, Kilimanjaro Archives (H1\22RBM-SIIP).

61. van Koppen, B., Sokile, C., Hatibu, N., Lankford, B., Mahoo, H., & Yanda, P. 2004. *Formal Water Rights in Tanzania: Deepening the Dichotomy?* Colombo: International Water Management Institute.

62. Andersson, Roger, Wänseth, Fritz, Cuellar, Melinda & v. Mitzlaff, Ulrike. 2006. *Pangani Falls Re-development Project in Tanzania*. Stockholm: Swedish International Development Cooperation Agency (SIDA), Department for Infrastructure and Economic Cooperation.

63. Among these are Kilimanjaro Joint Action Project (PAMOJA), Traditional Irrigation and Environmental Development Organisation (TIP).

64. TIP. "Water Use Management and Traditional Irrigation Systems in Tanzania: A presentation for the Second World Water Forum", World Water Fair, The Hague, 17–22 March 2000. TIP Archive, Moshi.

65. Hans, Keizer. 2001. "Catchment Management within the Pangani Basin", Workshop, 6–7 December 2000. TIP Archive, Moshi. H. Keizer personal communication, 2001; field conversations with TIP staff, Moshi, June 2003.

66. Zongolo, S.A. "Identification of problems experienced in irrigation practice in Runduguai area, Hai District", PAMOJA Memo, Moshi, December 2001.

67. United Republic of Tanzania, Proposed Draft Bill for the Water Resources Management Act.

68. United Republic of Tanzania. National Water Sector Development Strategy, 15.

69. United Republic of Tanzania. National Water Sector Development Strategy, 57.

70. Pangani Basin Water Office file PBWO/PBWOQR # 24: Quarterly Report, October-December 2004, p. 9. United Republic of Tanzania, Developing a Forum for Water Users in the Kikuletwa River Catchment. Moshi: Pangani River Basin Management Project, 2006.

71. Vavrus, Frances K. 2003. "A Shadow of the Real Thing: Furrow Societies, Water User Associations, and Democratic Practices in the Kilimanjaro Region, Tanzania" in *The Journal of African American History, 88*,pp. 393-402. Cleaver, Frances & Toner, Anna. 2006. "The Evolution of Community Water Governance in Uchira, Tanzania: The implications for Equality of Access, Sustainability and Effectiveness" *Natural Resources Forum 30,* No. 3, pp. 207–218.

72. For a related discussion, see Kemerink, J. S., Ahlers, R. & Van der Zaag, P. 2009. "Assessment of the Potential for Hydro-solidarity within Plural Legal Conditions of Traditional Irrigation Systems in Northern Tanzania" in *Physics and Chemistry of the Earth, parts A/B/C, 34,* pp. 881-889.

73. For further discussion of indigenous models of irrigation organisation, see Tagseth & Mattias, *Irrigation Amongst the Chagga in Kilimanjaro, Tanzania: The organisation of "Mfongo" Irrigation*.

74. United Republic of Tanzania, National Water Policy & National Water Sector Development Strategy.

75. The local government reforms and the reforms towards village land titling in Tanzania from the 1990s are two examples.

Trans-boundary River Basins

Hydropolitics in the Horn of Africa:
Conflicts and Cooperation in the Juba and Shabelle Rivers

Abdullahi Elmi Mohamed and Hussien M. Iman

Figure 1: Map of Somalia showing the Juba and Shabelle Rivers

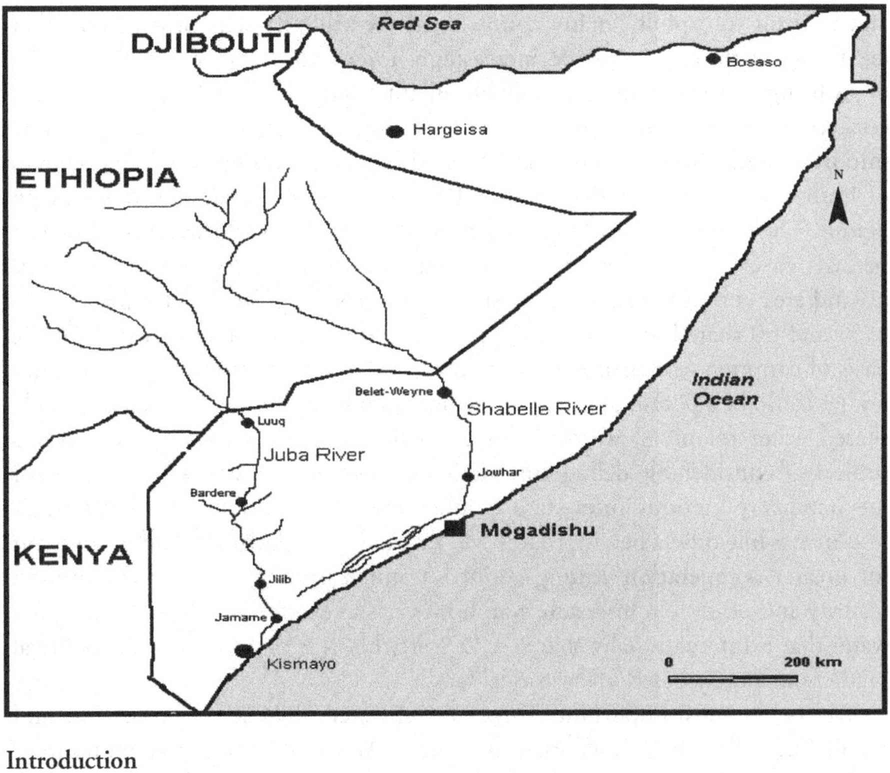

Introduction

Water – a basic human necessity is a critical resource for all aspects of human existence, environmental survival, economic development and good quality of life. Freshwater is one of the most essential of the elements that support human life and economic growth and development as the UN[1] identified lack of freshwater as being one of the five major problems facing humanity, while the UNEP[2] reported water shortage as one of the two most worrying problems for the new millennium. No other resource affects so many areas of economy or of human and environmental health.[3] The whole issue of global food security is closely linked to water availability[4]. Globally, freshwater constitute only 2.5% of all waters on the Earth and most of the easily-available freshwater resources exist in rivers and lakes[5] that are shared by several countries. This finite resource in the world is further limited by its uneven distributions in world's regions and countries, and its international status. Worldwide, millions of people do not have secure and safe access to freshwater and adequate sanitation, causing millions of death. Moreover, the increasing population puts greater demand on freshwater supplies. Globally, as far as water resources management is concerned, three related world trends strongly increase demand for water and greatly exacerbate the water situation. These are population growth, economic development and climate change.

In general terms, water resources management becomes increasingly critical and as new local and national sources of water become scarce, limited, expensive and difficult to exploit[6], many countries in the arid and semi-arid regions that are facing water crisis[7] will be increasingly forced to consider the possibilities of utilizing the water that is available in international river basins. Thus, the concerns relating to the use of international waters are becoming increasingly more important and complex because most remaining major easily-exploitable sources of freshwater are now in river basins that are shared by two or more sovereign States[8]. The number of international river basins in the world were identified to be 261, covering 45.3% of the world land area excluding Antarctica[9]. In terms of land area within international basins, Africa has the greatest percentage of all, 62% and 60 shared watercourses.[10] The lack of integrated management on the basis of common management for most of the continent's trans boundary water bodies could be a potential threat to regional stability. The issue of internationally shared water resources, which is highly political in its nature[11], is currently a subject of considerable debate internationally. Some have stressed the apparent inevitability of serious inter-state conflict over competition for shared water resources while others believe that it will provide an opportunity and instrument for greater co-operation among countries and reasons to search for common security and peace. No interstate war, however, has ever been fought for access to water that is internationally shared[12]. Despite this, it is difficult to make political predictions purely based on historical facts.

Rivers, the most important source of freshwater available for human use and the lifelines of many impoverished nations in Africa whose primary economy is agriculture, are increasingly coming under a lot of strain. In general, internationally shared rivers particularly those in dry climate regions could be a source of conflict or a reason for cooperation between countries sharing them.

Scope and Methodology

In the second half of the last century, it was observed that the concerns relating to the use of international water are becoming increasingly more important and complex. Water is a scarce resource in the Horn of African region, where the Juba and Shabelle River Basins are geographically located, as shown in figure 2 below.

Figure 2: Factors influencing Management of International River Water Resources

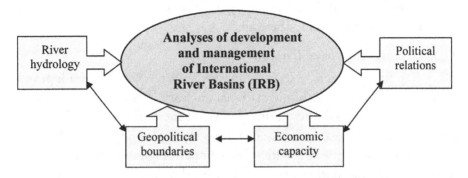

The overall purpose of this paper is to examine the physical and developmental aspects of the two rivers, the Juba and Shabelle in the Horn of Africa and to analyse the resulting hydropolitics and the looming water conflicts. The analysis will be based on the interrelated factors in Figure 2 assumed to have large influence on the joint management and development of shared water resources in International River Basins. The paper also presents some aspects of interaction between Somalia and Ethiopia over these common river systems. The paper suggests an alternative solution to prevent future water conflict over these shared resources in the two basins.

Information for this paper was collected through literature review, as well as information gathered from relevant organizations and other sources. The paper will first present the issue of international river basins from the point of view of international legal frameworks and institutional mechanisms. Interviews with relevant people on the issue of the effectiveness of the international water law were also carried out. Information gathered was presented and analysed in a manner aimed at describing the selected river basins from a physical, hydrological, economic as well as developmental aspect. This will be followed by a hydropolitical analysis of the two river basins.

International Rivers

According to a study carried out by a group of scientists,[13] the number of international river basins in the world were identified to be 261, covering 45.3% of the world land area excluding Antarctica. In terms of land area within international basins, Africa has the greatest percentage of all, 62%. Five of the world's eight rivers that pass through eight or more countries are in Africa.[14]

International Legal Perspectives on International River Basins

• The Application of Different Theories in International Water Law

The various uses to which the shared waters of international river basins are put create problems of both technical and juridical nature.[15] Problems of the latter category concern notably the sovereignty of the basin States.

Traditionally, international water law recognizes four main theories[16] that attempt to define and delineate the rights of basin states to use water from a shared river system. These are:

• The Theory of Absolute Territorial Sovereignty

According to this theory, a state, as 'master of its territory', may adopt in regard to watercourse within its national territory all measures deemed suitable to its national interest, irrespective of their effects beyond its borders.[17] In other words, a state can do as it pleases and is entitled to do as it chooses with its waters within its boundaries ignoring the effects of its actions on neighboring states. Obviously, this theory is favorable to upper basin state. Downstream states have of course always opposed the absolute territorial sovereignty doctrine, which is never implemented in any water treaty.

• The Theory of Absolute Territorial Integrity

This theory, which is also known as the theory of natural flow of river, is the direct opposite of that of absolute territorial sovereignty. It espouses the old common law doctrine of water rights whereby a lower riparian (basin state) claims the right to the continued, uninterrupted (or natural) flow and unaltered condition of the water from the territory of the upper riparian state.[18] According to this theory, a riparian state may not proceed with the harnessing of a section of an international waterway travers-ing its territory if it is of a nature to cause injury to the interests of other basin States. This theory is thus favourable to the lower-basin state(s) and awards 'a veritable right of veto to downstream states'.[19]

• The Theory of Limited Territorial Sovereignty and of Limited Territorial Integrity

These theories are in practice complementary and even identical there-fore, they can be considered together. They consist of the assertion that every state is free to use the waters flowing on its territory, on condition that such utilization in no way prejudices the territory or interests of other states. Permitting the use of rivers as far as no harm is done to other riparian States, these theories are where the concept of reasonable use originated.

• The Theory of Community of Interests in the Waters

In an attempt to advance and improve on the doctrine of limited ter-ritorial sovereignty and integrity, this theory insists on a 'community approach'. This doctrine suggests that all basin states have a common interest in developing the basin.[20] Under this theory, state boundaries should be ignored and drainage basin is regarded as an economic and geographic unit. This doctrine represents a more balanced approach that seeks to contribute to the joint development of riparian states within a shared basin through equitable division and sharing of benefits.

Each of these theories reflect different historical and judicial approaches to solving the problems experienced by riparian States[21] and also reflect an important change from rights to ownership of water, to one which strives to ensure that the interests of all parties are met equitably and efficiently. The international law, as an instrument of regulations on the trans-boundary freshwater issues, is at present inconclusive and weak.[22] Management of shared water resources in international river basins might be possible only if the affected and concerned countries accept the limitation to their sovereignty over the common water resources. It requires mutual agreement to define this limitation. This is, of course, the obstacle, which can hinder the development of a partnership between the riparian states.

Law is an instrument that can be used to smooth out conflicts of interest generated, for instance, in the sharing of water resources. However, the utilisation and management of a shared river is subject not only to man-made laws but also the natural laws, which do not recognize the political and geographical boundaries. In order to form a framework for cooperation, States sharing water need first to settle their inter-State dispute over the water in question on a legal basis. The absence of formal political agreements contributes to this problem. On the other hand, since each river basin is unique, with its own economic, geographical, ecological and political variables, no comprehensive system of rigid rules can anticipate adequately the variations from one basin to another.

Settlements of Disputes and Conflict Resolutions

When States disagree over the way in which water resources of an international river basin should be utilized, they have to resort to some dispute settlement procedures. Before parties or States sharing water system go into a conflict, which might develop from a water security problem to a national security problem,[23] there are three main procedures and methods used to find a solution to the problem. These are (1) negotiation and consultation; (2) mediation and conciliation; and (3) arbitration and adjudication. Disputes over shared water bodies can normally arise at least under two different circumstances; one can be riparian states that have never entered a treaty or agreement, while the other can be an existing agreement which can not provide a binding decision to the problem in dispute.

UN Convention on the Law of Non-navigational Uses of International Watercourses

In a move to reconcile conflicting theories favouring either upstream or downstream countries and resolve the alarming crises in shared freshwater resources, the UN General Assembly adopted the UN Convention on the Law of Non-navigational

Uses of International Watercourses in 1997. This Convention, which is not yet formally ratified and thus not operational, encourages cooperation in order to address equitable, reasonable and non-harmful utilization of international freshwaters. Many argue that this new legal instrument is too weak to guide mediation of basin states disputing over shared water resources, such as those in the Horn of Africa sharing the Juba and Shabelle rivers, where dispute has been experienced in the past and further disputes potentially exist.

The UN Convention is however an international framework instrument, which may apply in the absence of agreement over shared freshwaters. Cooperation through a joint mechanism in the form of commission is a vital necessity if the aims are to achieve equitable, reasonable and non-harmful utilization of the international watercourse. This is what the UN Convention encourages. The backbone of the UN Convention is informed by two principles: the principles of equitable and reasonable utilisation, and the principle obligation not to cause significant harm. The No-Harm rule forms a fundamental rule, while the principle of equitable and reasonable utilization is regarded as the cornerstone of international water law.

The No-Harm rule seems to favour the lower riparian States, which would benefit from the obligation posed on the upstream States not to cause a significant harm. This would automatically disadvantage the upper riparian States, as any use that causes harm would be deemed inequitable. Gleick[24] argues also that the inherent conflict between the equitable uses and the obligation not to cause harm is among the weaknesses of the UN Convention.

RIGHTS AND RESPONSIBILITIES OF UP- AND DOWN-STREAM RIPARIANS IN AN INTERNATIONAL RIVER SYSTEM

In terms of the UN Convention	Riparian Position	
	Upstream Riparian	Downstream Riparian
Rights	Equitable & reasonable utilization.	Not to be harmed.
Responsibilities	Cause no-harm	But allow equitable and reasonable utilization.

Source: Author's compilation.

The above table illustrates the controversy between the two most important articles of the UN Convention based on (i) the principle of equitable and reasonable utilization, and (ii) the obligation not to cause significant harm. As shown in the table, each state in any given river basin – whether upper or lower riparian – has both rights and responsibilities in terms of the UN Convention. The upstream riparian has the rights to equitable and reasonable use of the shared waters, while the downstream riparian has the responsibility to allow the common waters to be equitably shared. On the other hand, upstream riparian States are responsible for and obliged not to cause significant harm, while the downstream riparian States have the rights not to be significantly harmed. As it is unlikely that both can be achieved at the same time without problems, this proves to be a real dilemma in water sharing rules and allocation. However, it is obvious to note that any expanded or additional activities of water resources developments seem to be harmful to lower States, particularly in the arid and semi-arid regions, where water is naturally and increasingly becoming scarce. So, although the idea that 'rights' should be accompanied by 'obligation' gained acceptance, the UN Convention does not define any principles as being more superior and important than the other in the case of dispute over the utilization of the international watercourses. This analysis is easily applicable to the Juba and Shabelle river basins in the Horn of Africa.

Since the international water law that could solve the problem in legal terms still remains vague and uncertain, and many concepts and principles remain also unsettled[25], the allocation of the water resources in internationally shared river basins is increasingly becoming a major global challenge. The definition of the concepts, which are intentionally vague, guarantee continued ambiguity in the principles of customary law.[26] And there is some doubts raised over the viability of the UN Convention.[27] The principle of "reasonable and equitable utilisation" embodied in Article 5 of the UN Convention is somewhat vaguely worded, provides little guidance in this regard and is prone to subjective interpretations.[28] It has been argued that[29] the principles set forth in the UN Convention are not binding and offer little concrete guidance to the problem of allocation of water resources in international river basins. It is therefore doubtful whether the UN Convention will ever become truly operational.[30] It therefore seems impossible to form an institutional framework or arrangement that could serve the purpose of implementing the UN Convention because it is wide and vague. Many countries' governments are finding the Convention either too strong (upstream) or too weak (downstream). Utilization of shared water resources should not only be equitable but also sustainable. Therefore, the two concepts of "Equitable Use and Sustainable Development"[31] should be combined and harmonised.

There is no formally ratified rule of international water law. UN Convention is not yet operational, because the required number of countries did not yet ratify it. The UN Convention was open for signature until 21 May, 2000 and requires ratification by 35 States to enter into force and effect. Only sixteen countries have signed the UN Convention and about half of them ratified it.[32] The reluctance of the Member States to sign and ratify the Convention could be linked to the "no significant harm" component – and its relation to the equitable and reasonable utilization component. It may also be linked to what would be the impact of the UN Convention on existing as well as future international watercourse agreements. The most important concern of the UN Convention in its future is therefore the ratification by the UN member States.

It is therefore safe to say that that this new legal instrument is too weak to guide mediation of basin states disputing over shared water resources, such as those in the Horn of Africa sharing the Juba and Shabelle rivers, where dispute has been experienced in the past and further disputes potentially exist.

Physical Aspects of the Shabelle and Juba Rivers

Ethiopia, Kenya and Somalia occupy parts of the Juba and Shabelle River Basins in the Horn of Africa. Contrary to previous estimations,[33] the total drainage area of the two basins was recently estimated at 805, 100 sq. km.[34]

The Shabelle River Basin

Running a distance of about 1500 km, the Shabelle rises in the Ethiopian Highlands, where annual rainfall exceeds 1000 mm. Flowing generally south-easterly direction; the Shabelle River passes through an arid land in the eastern region of Ethiopia[35] cutting wide valleys in southern Somalia. The river does not normally enter the Indian Ocean, but into a depression area, where it is finally lost in the sand in southern Somalia. Only with exceptionally heavy rains does the Shabelle River break through to join the Juba and thus succeed in reaching the ocean. With an average annual rainfall of 455 mm and much higher potential evaporation, mean annual runoff of the Shabelle River at Belet-Weyne is 2,384 million m3.[36] Over 90% of the runoff is generated by catchments within Ethiopia.[37] As the river crosses the existing international border between Ethiopia and Somalia, the Somali City of Belet-Weyne in the Hiiraan region is the most important point where the river flow and its water quality could be observed inside Somalia. The river has a high saline content even during high flows.

The Juba River Basin

Like the Shabelle, the Juba River originates from the Ethiopian Highlands, where three large tributaries: the Genale, the Dawa and the Weyb meet near the border with Somalia to form what is known as the Juba River inside Somalia. Rainfall at the source reaches 1500 mm/y, dramatically decreasing southwards, where the mean reduces to 550 mm in Somalia. Luuq, a Somali town, is the most important point to observe the runoff of the Juba River as it crosses the border. The Juba, which enters the Indian Ocean at Kismayo City, has a total length of 1100 km, 550 km of which in Somalia. The mean annual runoff at Luuq is 6,400 million m³. Kenya is part of the basin but as there are no tributaries originating there and significant flow of surface runoff, it does not normally contribute to the Juba, and has no access to the main river, thus has no significant interests.

The Shabelle is larger in size and longer in distance than the Juba, but this does not lead the Shabelle to be larger in runoff due to climatic and geological conditions. As Somalia's most water resources exist in these rivers, runoff contributions by catchments in Somalia are normally minimal. The effects of this factor will be analysed later in the paper.

Developmental Aspects of the Shabelle and Juba Rivers

In upstream areas of Ethiopia, there are few developments based on the two rivers' water resources. In 1988, Ethiopia completed the Melka Wakana hydroelectric project on the upper reaches of the Shabelle. Ethiopia has carried out a master plan study in the Shabelle river basin. As an outcome from that master plan study, Ethiopia constructed a major dam on the Shabelle in 2001 near Godey for irrigation development. Due to the very narrow arable alluvial plains, there are few permanent agricultural settlements along the Shabelle River in the Somali National Region of Ethiopia. The majority of these people living in the region are traditionally nomads in semi-desert environment.

Ethiopia is now carrying out a master plan study on the Genale-Dawa rivers sub-basin which is part of all the river basins of the country on the basis of which the sub-basin development plan will be formulated. The specific objectives of the Genale-Dawa River Basin Integrated Resources Development Master Plan Study are to study the most attractive water resources projects in the sub-basin. These water projects, when undertaken, will have significant impact on the downstream uses. The three year study period is financed by the African Development Bank.

As the two rivers supply the Somalia's rice bowl and support important economic areas in southern Somalia, several agricultural development projects have been implemented based on the water resources of the two rivers. Irrigation projects that were implemented or planned on the Juba River include: Juba Sugar Project (JSP), often known as Mareerey, irrigating sugarcane near Jilib; Mugaambo Rice Irrigation Project near Jamame, using run-off from the river via canals; Fanole Dam Project, multipurpose dam development for irrigation, hydropower generation and flood mitigation, located near Jilib; Arare Banana Irrigation Project, Jamame; Bardere Dam Project (BDP), the largest ever planned but unimplemented development project, which will be discussed below.

Inside Somalia, no major dam development was built on the Shabelle River, but there are many agricultural activities along the Shabelle River that intensively use much of the available water. An off-stream facility with a storage capacity of 200 million cubic meters was built near Jowhar in central Somalia. Another dam which will store 130 - 200 million m^3 was proposed to be built upstream of Jowhar. Several agricultural activities exist in areas around Mogadishu.

Hydropolitical Aspects of the Shabelle and Juba Rivers

Hydropolitical history between Somalia and Ethiopia has always been contentious due to territorial conflicts (the Ogaden disputed region) and the lack of coordination in planning development projects.

Historical Conflicts and Current Tensions

The relations between the two states of Ethiopia and Somalia were complicated particularly in view of their long history, which is full of animosity, mistrust, conflict and border disputes, which resulted from the demarcations by the European Colonialists during the 19th and 20th centuries. During that period, Ethiopia played a key role in the colonial division of the Somali Plateau into five areas. These tense relations resulted in political conflict which caused at least two military wars in 1964 and 1977.

The relations have also been deteriorating since the overthrow of the two countries' dictators in 1991. Since 1996, Ethiopia has been criticized severally for its repeated military and political interventions in Somalia, a country lacking a central government since 1991. In August 2000, when the Ethiopian Prime Minister attended the inauguration of the rebirth of the Somali Government, many people looked upon it as a new era for Ethiopia-Somali relations, but this hope has been continuously dashed ever since. The transitional national government of Somalia (TNG), resulting from the peace process sponsored by Djibouti, tried a number of times, with no encouraging results, to normalize the uneasy relations between Somalia and Ethiopia. In the ongoing international war against terrorism led by USA, the Ethiopian Government officially said that there are terrorist groups linked to Al-Qaeda Network inside Somalia, which the TNG strongly denied. It is certain that these unfavourable relations will adversely affect future cooperation for the development of these shared rivers. The two countries have in the past never discussed agreements nor joint commission for the utilization of the shared rivers.

Shabelle Development Projects in Ethiopia

During the 1950s, there was a large scale *Shabelle Development Scheme* planned in Ethiopia, which was not implemented. Ethiopian plans in late 1970s towards development of the Shabelle River in most upstream areas for irrigation concerned Somalia. Resulted from its national policy of food self-sufficiency, Ethiopia has, since 1991 after the overthrow of the miltary regime, gone into a new process of planning for water resources development. Taking advantage of Somalia's deep political crisis, Ethiopia started building large dams on the Shabelle River. According to WIC[38], the major dam built by Ethiopia in 2001 near Godey for irrigation purpose would reduce the river runoff by more than 60%. The potential impacts of this dam development on the downstream use as well as on the natural environment are substantial, but there is no voice heard from the affected communities so far.

Existing and planned dams on the Shabelle River in Ethiopia function also as a political weapon for its rival downstream riparian. As many activities in southern Somalia, where the two rivers enter and supply, depend mainly on this river's water resources, unilateral developments that Ethiopia currently carries out will severely impact on Somalia both in terms of economy and environment. Actions reflect and imply existing policies and perhaps the unilateral Ethiopian actions are based on its previous argument saying that it is the sovereign right of any riparian state, in the absence of an international agreement, to proceed unilaterally with the development of shared water resources within its territory. These new Ethiopian dams on the Shabelle will exacerbate the silent border dispute between the two countries.

Juba Valley Development in Somalia and Ethiopian Plans

The need to regulate the Juba River was recognized as early as the 1920s by the Italian colonial administration in Somalia. Since then and particularly after independence in 1960, the Juba and Shabelle valleys became the focus of country's economic development. The largest ever-planned water development project was the Bardere Dam Project (BDP) launched during the 1980s on the Juba River near the town of Bardere that fully utilized the river water. Regarded as a vital step towards food self-sufficiency, which received priority in development planning, the BDP is intended for flood mitigation, irrigation development and hydropower generation. It would irrigate about 175,000 hectares of agricultural land and supply hydropower to reduce the cost of petroleum imports. The BDP was economically and technically motivated but failed politically. The two political factors that played an important role in its failure were: (1) the dictatorial regime which Somalia had at the time of the project appraisal and the deteriorating political situation of the country during the 1980s which resulted in the ongoing civil war and became a major hinderance to the project development. The civil war which erupted in 1991 interrupted and derailed the entire project; (2) strong opposition from upstream co-basin country of Ethiopia impacted the project as it argued that the river crosses disputed land and has no agreement on the utilization of its waters. Because of the Ethiopian opposition, the size of the dam has been substantially reduced to irrigate only 50,000 hectares.

Currently, Ethiopia is in the process of developing the basin's water resources unilateraly. The ongoing master plan study in the Genale-Dawa rivers of the Juba Basin which is carried out to formulate water development projects will have major significant impact on the downstream communities both in terms of water availability and flood risk.

The Role of the Rivers in Somalia's Economic Development

The Juba and Shabelle Rivers are an important resource base for Somalia, but there are growing fears that these rivers may impoverish the nation they would set on the path to prosperity because of water scarcity and upstream activities. Somalia lacks significant alternatives to the two rivers as long as water developments for agricultural productions are concerned. Ongoing as well as traditional socioeconomic activities in southern Somalia are heavily reliant on the availability of water in the two rivers, and without the guaranteed access to water the fertile areas between the rivers would have no value. Water resources in the two rivers are strongly linked to the survival of the Somali national economy as well as its social and environmental well-being, thus the security of the nation.

47

Institutional structures and capacity for management of water affairs are totally absent in Somalia. Water infrastructure that was set up for irrigation was also destroyed during the civil war.

Growing Water Scarcity and Looming Water Conflict

Considering the possible and potential future water development plans and taking into account the limited amount of water, the water resource of the two rivers is unlikely to fulfil the sum of all demands by the basin countries in the future. Potential disputes over the shared rivers are therefore likely to arise in response to political instability and desire for economic development. This may result in competition over the utilization of the scarce water in the rivers, which together with the current and historical relations between the two basin countries, may lead to an international conflict, thereby shifting the problem from a water sharing to a national security issue.

While it was previously assumed that trans-boundary water was a source of conflict and even war, research suggests that states with a shared water body in almost all cases choose to cooperate over the resource even in the event of other political conflicts. But this is not the case of the riparians of the Juba and Shabelle river basins in the Horn of Africa. According to Clarke,[39] Ethiopia argued in the Nile case that it is "the sovereign right of any riparian state, in the absence of an international agreement, to proceed unilaterally with the development of water resources within its territory". The same argument may be applied to the Juba and Shabelle rivers issue due to the absence of an international agreement between Ethiopia and Somalia on these shared rivers.

However, the factors that may increase the risk of future water conflict between the riparian states of the Juba and Shabelle river basins are many. These include severity of the water scarcity in the riparian countries; the effects of climate change on the region; geographical balance of the problem; historical conflicts and current misunderstandings; relative economic strength and military power, and a growing population.

Lessons from the Nile River Basin

Physical Geography

The Nile River Basin, with an area of about 3 million km², is geographically shared by ten countries in Africa: Burundi, Democratic Republic of Congo, Egypt, Eritrea, Ethiopia, Kenya, Rwanda, Sudan, Tanzania and Uganda. The Nile River is the longest river in the world with a length estimated to be 6,825 km from its source in Lake Victoria to the Mediterranean Sea. The basin encompasses two major river systems or sub-basins namely the White Nile and the Blue Nile. The principle source of the White Nile is the Lake Victoria, while the Blue Nile has its principle source in Lake Tana in Ethiopia. The Blue Nile joins the White Nile in the Sudanese capital, Khartoum. The average annual flow at Aswan in Egypt is about 84 billion m³, which is ten times more than the combined annual flows of the Juba and the Shabelle rivers. Nile waters come from rainfall on the Ethiopian highlands and the catchment areas of the equatorial lakes. The northern part of the basin has virtually no rainfall in the summer, while the southern area has heavy rains during the summer months. Its ecological system is unique, hosting a number of varied landscapes, with high mountains, tropical forests, woodlands, lakes, savannas, wetlands, arid lands, and deserts.

Economic Geography

Agriculture is the primary economic activity in all riparian countries of the Nile Basin. In Egypt and Sudan, irrigated agriculture is the dominant sector. Over 5.5 million hectares are under irrigation, with plans to further irrigate an area of over 4.9 million hectares. The present irrigation in the upstream White Nile riparian areas is very small and there are plans for a future expansion over an area of 387,000 hectares in Uganda, Tanzania and Kenya. In Ethiopia, the potential identified in the Blue Nile basin includes 100,000 hectares of perennial irrigation and 165,000 hectares of small-scale seasonal irrigation. The other riparian countries have no potential for irrigation in the basin and depend almost completely on rain-fed agriculture. At present, roughly 300 million live within the ten countries that share the Nile waters, and it is expected to grow to 591 million by 2025 at an average rate of 2.5 – 3.0%. Most of the countries of the Nile Basin lack the basic water infrastructure and human resources that are usually needed to implement and maintain large development projects.

Hydro-political Geography

The only operational agreement that is currently available in the use of the water resources of the Nile River is that between Egypt and Sudan signed in 1959 for full utilisation of the Nile waters. That agreement was only based on the needs of Egypt and Sudan at the time. Other riparian countries refused to accept the 1959 agreement, which made it possible for Egypt and Sudan to undertake a number of water development projects such as the Aswan High Dam in Egypt. When this agreement was signed in 1959, the UN Convention on the Law of Non-navigational Uses of International Watercourses was not in place.

The Nile Basin riparian countries agreed in 1999 to establish the Nile Basin Initiative (NBI) to provide an agreed basin-wide framework for the utilisation of the Nile River Basin resources to fight poverty and promote socio-economic development in the ten Nile countries. The NBI is an intergovernmental transitional arrangement that seeks to address the basin problems while a permanent agreement is worked out. The NBI is led by a Council of Ministers in charge of Water Affairs from the member states, and is supported by a Secretariat based in Uganda. Despite the vast but as yet only partially exploited developmental potential of the Nile's waters, the highly volatile political geography of the basin and of the individual states sharing the waters indicates that things will not improve anytime soon.[40]

One of the most important lessons that could be learned from the Nile river process is that any agreement on international rivers often takes long to be reached at. In addition to this, the current situation in Somalia and the hostile relation between the two countries seem to be an impediment to any inter-state cooperation on water-sharing between the riparian countries.

Conclusion

In both the Juba and Shabelle basins, Somalia is a vulnerable end-user located in the downstream area, which is the least favourable position to be in hydro-political terms, as the upstream basin country, Ethiopia, can theoretically divert and pollute the water in the rivers. This renders Somalia permanently and heavily dependent on the actions taken by Ethiopia. Consequently, the downstream users in Somalia are the hostages of upstream activities in Ethiopia. Although the issue of the Juba and Shabelle Rivers is a hidden and powerful one that could explode at any time in the future, no negotiations could be initiated before addressing and solving other more fundamental causes of the historical conflicts and the current tensions. In view of the region's current political conditions as well as the historical facts combined with the future desire to increase the utilisation of the available resources in the river basins, it is unlikely that the desperately needed cooperation will be realised and future water conflict seems to be inevitable. It may also turn out to be another mid-century international conflict, if nothing is done.

As these shared waters play a key role in future relations between Ethiopia and Somalia, there is a great need to initiate cooperation, through dialogue based on mutual security, in order to establish significant trust. The only assurance that no harm will be done to the interests of any party lies in the process of collaboration through negotiation. A useful way to initiate and sustain dialogue is to seek opportunities for mutual benefits. One opportunity, demanding great political commitment, that could be explored is to go into regional economic integration based on water through securing a reliable access to the sea which Ethiopia desperately needs in exchange to undisturbed river flows for Somalia. Since Eritrea's independence from Ethiopia in 1993, Ethiopia was left in a precarious situation regarding its lack of a reliable outlet access to the sea for the economy. Perhaps, in view of this, the existing opportunity, from which the two countries, Ethiopia and Somalia could mutually benefit, is to allow the two rivers to run into Somalia without any major developments being implemented in upstream Ethiopia in exchange for freely accessible ports for Ethiopia along Somalia's long coastal lines. This economic integration demands strong political commitment from the two countries to ensure a joint security and safe co-existence in the future.

Notes

1. UN 'UN Panel of Futurologists', 1998. *Expert Panel on Problems Facing Humanity in the Next 25 Years*. New York.

2. UNEP 'United Nations Environment Program', 1999. *Everybody Lives Downstream*. Press Release for World Water Day 1999.

3. Feder, Gershon & Le Moigne, Guy, 1994. "Managing Water in a Sustainable Manner": *Finance & Development, vol. 31*, no. 2, June 1994.

4. Falkenmark, Malin, 1997. Meeting Water Requirement of an Expanding World. *Philosophical Transactions of the Royal Society Biological Sceinces. vol. 352*. The Royal Soceity, London. pp. 929.

5. Shiklomanov, I., A., 1997. *Assessment of Water Resources and Water Availability in the World*. Background Report to Comprehensive Assessment of Freshwater Resources of the World (UN/SEI, 1997).

6. Delli Priscoli, Jerome, 1998. *Water and Civilization: Conflict, cooperation and the roots of a new eco-realism*. Proceedings of the Eighth Stockholm World Water Symposium, 10-13 August 1998, Stockholm, Sweden. 17 pp.

7. Biswas, Asit, K., (ed.) 1996. *Water Resources Management: Environmental Planning, Management, and Development*. McGraw-Hill. p.185

8. Biswas, Asit, K., 1981. *"Water for the Third World", Foreign Affairs 26, no. 1*, pp.144-166 and World Bank, 1998. *International Watercourses: Enhancing Cooperation and Managing Conflict*, Salman M. A. Salman and Laurence Boisson de Chazournes.

9. Wolf Aaron, T., Natharius, J. A., Danielson, J., J., Ward, B., S., Pender, J., 1999. International River Basins of the World. *International Journal of Water Resources Development, Vol. 15*, No. 4.

10. *International Journal of Water Resources Development, Vol. 15, No. 4.*.

11. Nakayama, Mikiyasu, 1999. *Politics Behind Zambezi Action Plan*. Water Policy 1-13.

12. Wolf, Aaron, 1997. *"Water War" and "Water Reality": Conflict and Cooperation Along International Waterways*. Presented at the NATO Advanced Research Workshop on Environment Change, Adaptation and Human Security, Budapest, Hungary, 9-12 October, 1997. p.11 and Delli Priscoli, Jerome, 1998. *Water and civilization: Conflict, cooperation and the roots of a new eco-realism*. Proceedings of the Eighth Stockholm World Water Symposium, 10-13 August 1998, Stockholm, Sweden. 17 pp.

13. Wolf Aaron, T., Natharius, J. A., Danielson, J., J., Ward, B., S., Pender, J., 1999. International River Basins of the World. *International Journal of Water Resources Development, Vol. 15*, No. 4.

14. *International Journal of Water Resources Development, Vol.* 15, No. 4.

15. Godana, Bonaya, Adhie, 1985. *Africa's Shared Water Resources: Legal and Institutional Aspects of the Nile, Niger and Senegal River Systems*. London : Pinter; p.11, 21.

16. Utton, A. E. and Teclaff, L., 1978. *Water in a Developing World: The Management of a Critical Resource, Western Special Studies in Natural Resources and Energy Management*. United Nations Development Programme, New York.

17. Menon, P.K., 1975. "Water Resources Development of International Rivers with Special Reference to the Developing World", *International Lawyers, Vol. 9*, pp.441-64.

18. Godana, Bonaya, Adhie, 1985. *Africa's Shared Water Resources: Legal and Institutional Aspects of the Nile, Niger and Senegal River Systems*. London : Pinter; p.11, 21

19. *Africa's Shared Water Resources: Legal and Institutional Aspects of the Nile, Niger and Senegal River Systems*.

20. *Africa's Shared Water Resources: Legal and Institutional Aspects of the Nile, Niger and Senegal River Systems*.

21. International Law Association (ILA) 1966. Report of the Committee on the Uses of the Water on International Rivers. London, 1994.

22. Naff, Thomas. 1993 http://en.scientificcommons.org/57170082.

23. Carlson, Ingvar,1999. National Sovereignty and International Watercourses. High level Panel Debate at the 9th Stockholm Water Symposium (SIWI), Stockholm. Mr. Carlson was Prime Minister of Sweden from1986 to1995.

24. Gleick, H., Peter, 1998. *The World's Water: The Biennial Report on Freshwater Resources, 1998-1999.* Pacific Institute for Studies in Development, Environment and Security. Oakland, California. Islands Press. p. 158, 210.

25. Caflisch, Lucius, 1998. *Regulation of the Uses of International Watercourses.* In International Watercourses: Enhancing Cooperation and Managing Conflict. Proceedings of a World Bank Seminar. World Bank technical paper no. 414. p.9, 16.

26. Wolf, Aaron, 1999. Criteria for equitable allocation: The heart of International Water Conflict. *Natural Resources Forum, vol. 23*, no. 1. p.4, 6, 7.

27. Caflisch, Lucius, 1998. Regulation of the Uses of International Watercourses. In International Watercourses: Enhancing Cooperation and Managing Conflict. Proceedings of a World Bank Seminar. World Bank technical paper no. 414. p.9, 16.

28. Van der Zaag, P., I.M. Seyam & H.H.G. Savenije, 2000. *"Towards objective criteria for the equitable sharing of international water resources".* Proceedings of the International Association of Hydraulic Research, Windhoek, Namibia, 7-9 June 2000.

29. Gleick, H., Peter, 1998. *The World's Water: The Biennial Report on Freshwater Resources, 1998-1999.* Pacific Institute for Studies in Development, Environment and Security. Oakland, California. Islands Press. p. 158, 210.

30. Caflisch, Lucius, 1998. *Regulation of the Uses of International Watercourses.* In International Watercourses: Enhancing Cooperation and Managing Conflict. Proceedings of a World Bank Seminar. World Bank technical paper no. 414. p.9, 16.

31. Sustainable water resources management is a concept that emphazises the need to consider the long-term future as well as the present. Sustainable water resources systems are those designed and managed to fully contribute to the objectives of society, now and in the future, while maintaning their ecological, environmental, and hydrological integrity (ASCE, 1998; UNESCO, 1999).

32. Salman M., A., Salman, 2001. International Rivers as Boundaries - The Dispute over the Kasikili/Sedudu Island and the Decision of the International Court of Justice. *Water International, Volume 25,* No. 4.

33. National Water Center, 1989. The Shabelle River Basin, shared by Ethiopia and Somalia, is about 307,000 km², with more than half it in Ethiopia, while the Juba River Basin is 233,000 km²; with 65% in Ethiopia, 30% in Somalia and 5% in Kenya.

34. A study updating international rivers of the world (Wolf et al., 1999), gives the combined area of the Juba and Shabelle Rivers Basins as 805, 100 km², of which Ethiopia occupies 45.7%, Somalia 27.5% and Kenya 26.8%.

35. Inhabitants of the eastern part of Ethiopia are ethnically Somalis. This region was internationally known as Ogaden but in Somalia it is referred to as Somali Western. Recently, it has been named as Region 5 in Ethiopia.

36. United Nation's Food and Agriculture Organization (FAO), 1989. *A Brief Description of Major Drainage Basins Affecting Somalia.* Prepared by D. Kammer. National Water Centre, Mogadishu. Project Field Document No. 14. SOM/85/008.

37. National Water Centre, Mogadishu. Project Field Document No. 14. SOM/85/008.

38. Walta Information Center (WIC), 2001. An irrigation project worth 160 million birr is nearing completion in Godey, Addis Ababa, Ethiopia.

39. Clarke, Robin, 1991. *Water: The International Crisis.* London: Earthscan Publications. p. 104.

40. Elhance, P. Arun, 1999. Hydropolitics in the 3rd World: Conflict and Cooperation in International River Basins. United States Institute of Peace Press, Washington.

Nile Basin Initiative:
A Possibility of Turning Conflicts into Opportunities

C.A. Mumma Martinon, PhD

Introduction to the Nile Basin

River Nile, the longest in the world, is a vital source of water for millions of people in the North Eastern region of Africa. It flows from south to north, 6,825 kilometres over 35 degrees of latitude.[1] The Nile catchment basin covers approximately a tenth of the African continent, with an area of 3,007,000 square kilometres.[2]

The river has two main tributaries: the White Nile originating from East African highlands and the Blue Nile in Ethiopia. At Khartoum, the Blue Nile and the White Nile merge into one River Nile, 320 km north of Khartoum where the Atbara River, that rises in the Ethiopian Highlands, joins it.

The Nile waters are mainly from precipitation in the Ethiopian plateau and the mountainous hinterland of the Great Lakes.[3] Nearly one-third of Lake Victoria's entire inflow is derived from the 60,000 km² Kagera River catchment, while 33% and 10% of the catchment area fall within Tanzania and Uganda[4] respectively. The Nile receives no additional water during the rest of its 3,000 km journey through the desert to the Mediterranean Sea.

The total population of the Nile basin countries was 287 million people in 1995, and it is estimated that over half of it depends on the Nile waters.[5] By the year 2025, the total population of the basin is estimated to reach 597 million.[6]

Existing Conflicts amongst the Nile Basin Countries

Divergent Interests, Perceptions and Expectations

The ten Nile Basin riparian states have divergent interests, perceptions, and expectations of the Nile Basin Inititative (NBI) co-operation. Furthermore, the water resources in this region are unevenly distributed and used.

Egypt, Ethiopia, and the Nile

The Nile River represents and means different things to Egypt and Ethiopia. Egypt is totally dependent on the Nile waters. Consequently, the Nile is perceived as Egypt's lifeline. This is described best by the repeatedly quoted statement, "Egypt is the gift of the Nile", uttered by the Greek Historian Herodotus in 460 BC. "If the waters of the Nile have meant life for Egypt, then, they have meant something different for the Ethiopians."[7]

Figure 1: Map of the Nile Basin Countries

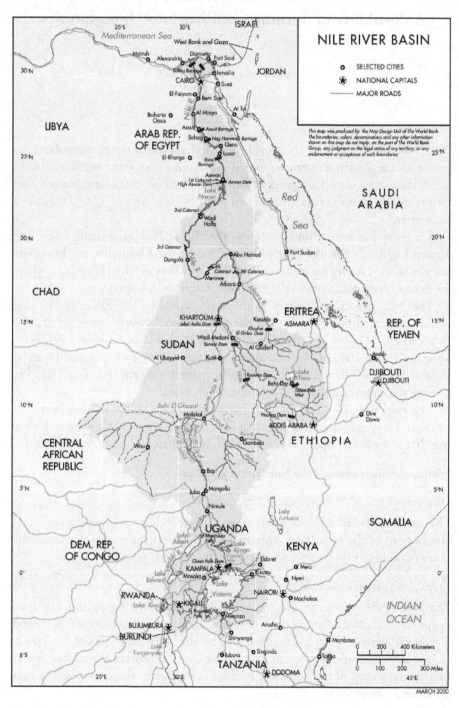

The NBI is generally viewed as a positive development. Ethiopia, however, is sceptical of Egypt's intentions and questions its commitment to genuine basin-wide co-operation. Egypt's alleged involvement (the Ethiopian government's allegation) in the Ethiopia-Eritrean conflict (in 1998-2000) by supporting the Eritrean regime is pointed out as an example - claims the Egyptian authorities deny.[8]

Ethiopia generally argues that the new co-operation efforts will become meaningful and effective if, and only if, the 1959 bilateral agreement between the Sudan and Egypt is nullified and a fresh Nile water redistribution arrangement that would accommodate the interests of all the riparian countries is negotiated.

Egypt on the other hand, underlines the fact that it is heavily dependent on the Nile waters, and it is thus a national security issue. They also emphasise the importance of basin-wide cooperation; stressing the urgency of such cooperation in the face of the serious challenges that the Nile basin countries are faced with while the issue of sharing Nile waters is down-played.[9]

Eritrea, which officially became an independent country in 1993, is a new state in the region, and joined the evolving Nile basin co-operation as an observer. The country up to now has chosen to maintain its observer status. Eritrea is interested in development for irrigation and hydropower of two seasonal streams that flow from its territory into the Sudan. The Mereb and Tekeze rivers form part of the border between Eritrea and Ethiopia at different sections, but these rivers have neither been the cause nor object of dispute during the recent fighting between the two countries.[10]

Sudan's Nile Priorities

The Sudan, with abundant surface water resources and arable land that could easily be brought under irrigation, has tremendous potential to become a major agricultural producer and exporter in the Nile basin.[11] The country is, thus, mainly concerned with development of its agricultural potential. Its ability to realise such potential, however, is limited by civil war in the South, economic crises, and the 1959 agreement with Egypt, which restricts how much of Nile water Sudan can use.[12]

Waterbury, who examines the major conflict of interests in the Nile basin involving Egypt, Ethiopia and the Sudan, identifies the Sudan as the "Master of the middle", referring to the vital position it occupies.[13] At the same time, the Sudan finds itself in a difficult position in that it is the only country that has signed the 1959 treaty on utilisation of the Nile water with Egypt and is generally a champion of this agreement when it comes to sharing the Nile waters with other riparian countries.

Great Lakes Region Countries

The Nile's riparians in the Great Lakes Region were unable to secure their interests in the Nile basin during the colonial era due to treaties signed on their behalf, which mainly favoured downstream riparians.[14]

It seems, however, that things have changed since then, as there are indications that many of the Great Lakes riparians are showing increasing interest in the Nile waters in order to promote their own development endeavours. In late 2002, for instance, the issue of the 1929 treaty which restricted the upper Nile countries from using the waters of Lake Victoria, and which gave Egypt advantages, was raised in both the Kenyan and the Ugandan parliaments. Members of Parliament in Kenya questioned the legality of the treaty, and called on the government to denounce it and seek support from other East African countries like Tanzania and Uganda[15]. Similar concern was voiced in the Ugandan parliament and it was further suggested that Egypt should pay an annual compensation to Uganda for use of the water in case of any new agreement in future.[16]

The Democratic Republic of Congo (DRC) has interest in shipping and fishing rights on Lake Albert, which is part of the Nile basin and forms part of the border between Congo and Uganda. Burundi and Rwanda are both members of the Kagera Basin Organisation (KBO), (the Kagera River rises on Burundi's territory), and enjoy high and regular rainfall. Their interest in the Kagera is confined mainly to hydropower generation. Tanzania is a member of the Lake Victoria Basin group as well as of the KBO.

The Nile Treaties as a Source of Conflict

To begin with, there is no binding international law on water utilisation.[17] The question of sharing water between upstream and downstream countries is still generally unresolved and an all too common source of conflict.[18] Biswas commented on this aspect by asserting that "the development and the management of international water bodies is a major challenge that will face us in the coming decades; it has to be admitted that the law on international waters is one of the most unsettled areas of international law." [19]

When it comes to the Nile *per se*, there is no multi-lateral and comprehensive agreement regarding the use of its water for the common benefit of all the people inhabiting the basin[20]. In this region, the most advanced bilateral agreements had remained partial instead of being comprehensive treaties. In a situation like this (no agreement), all national development plans made by one government concerning the use of the Nile will be perceived by another government as a threat to its national interest. Thus, they are a source of international conflict.[21] Many agreements have been made between these countries from as early as 15 April, 1891 – 8 November, 1959. All these legal instruments were negotiated on a strictly bilateral basis. The main bone of contention has been with the 1929 and 1959 treaties.

The 1929 Nile Agreement

This Agreement between Egypt and Great Britain gave Egypt "full utilisation of the Nile waters"[22] and simultaneously granted her the right to inspect and investigate any control work along the entire Nile. This agreement could not have had any binding effect on the Upper Nile riparian states for two reasons: (a) it was a bilateral agreement and did not include any of the upper riparian states; (b) it was struck by a colonial power, and this was even before the other countries got their independence.

The 1959 Nile Agreement

After Sudan gained its independence in 1956 it stated that "the 1929 agreement was no longer valid because it had been reached by Britain and Egypt without any consultation with the Sudanese and had discriminated against them."[23] After numerous discussions at ministerial levels on the subject, a deal on the full utilisation of the Nile waters was struck between Egypt and Sudan on 8 November, 1959. This agreement apportioned the total annual discharge of the main Nile, as measured at Aswan, i.e. 74 billion m^3; where Egypt was allotted 55.5 billion m^3, while the Sudan was allowed to use the remaining 18.5 billion m^3.

No room whatsoever was given to the other riparian countries including Ethiopia, which is contributing 85% of the total annual flow. Despite the fact that Ethiopia had got its independence at that time, it was not consulted or even invited to the negotiation process that led to the 1959 Agreement. While Egypt and Sudan contend that the 1959 treaty is valid, the upper countries led by Ethiopia reject it, and maintain that they have a right to exploit water resources within their borders. The countries have expressly stated, at different occasions, that they are not bound by the agreement on the basis of the cardinal principle in the law of treaties: *res inter alios acta*.

Attempts towards Managing the Nile Water Conflicts

Previous Attempts

After the independence of the upper Nile countries, other institutional frameworks had to be formulated to regulate the use, distribution, and management of the Nile waters. For the last three decades, various Nile countries have come up with different efforts to try and find solutions to these conflicts. These include The Hydromet Survey Project, Kagera River Basin Organisation, Undugu Tecconile and the former East African Community. However, in one way or the other, these co-operative ventures have not been successful.

These failed attempts implied that the potential of violent conflicts amongst these countries over water also increased. As a result, the Nile basin countries came up with further co-operative ventures. These included: Lake Victoria Environmental Management Project (LVEMP), Lake Victoria Fisheries Organisation (LVFO), East African Community (EAC) and Nile Basin Initiative (NBI), the latest initiative and the main focus of this paper. The initiatives before the NBI faced major problems when it came to solving the conflicts in the Nile region since they did not include all the ten countries of the Nile.

The Nile Basin Initiative (NBI): A Possibility of Turning Conflicts into Opportunities

The following makes the NBI different from the previous attempts: for the first time in history, nine out of ten countries of the Nile are represented within the NBI and the nine cover roughly a third of Africa's population. As an institutional framework, the NBI is meant to play an intermediatary role amongst these countries especially in areas where there are contradictory and opposing interests of the upper and lower countries, the NBI countries have come up with different activities and projects to create dialogue and better working relations amongst them.

In 1998, all the riparian states, except Eritrea, began discussions with a view to creating a regional partnership to better manage the Nile. A transitional mechanism for co-operation was officially launched in February 1999, in Dar es Salaam, Tanzania by the Council of Ministers of Water Affairs of the Nile Basin States (Nile-COM). In same year, the Nile-COM held a meeting in Addis Ababa under the chairmanship of Ethiopia and established the Nile Basin Initiative and later in the same year, a Secretariat was established in Entebbe, Uganda.

The NBI comprises of the following: establishment of a Shared Vision for the Nile which seeks "to achieve sustainable socio-economic development through the equitable utilisation of the benefits of the common Nile Basin Water Resources"; a *"Subsidiary Action Program"* to plan and implement actual investment projects at the country and sub-basin levels.

The guiding principle of the NBI is to look beyond national boundaries into ways of optimising and sharing equitably the benefits available to all through better water management, allocation and use. What this translates to is that the food – water linkages at national level should then be looked at internationally by all the countries of the Nile. Policy decisions about irrigation need to be made within a much broader socio – economic and political context. If the co-operation among the Nile countries has to be meaningful in any way, these are issues, which are obligatory and should be discussed and resolved within the co-operation. Important policy questions arise in this regard: In a water scarce and rainfall dependent region, would national irrigation expansion be the best approach for each country? Are there other options for the Nile countries to increase their food production rather than irrigation? Do mechanisms exist within the NBI which can enhance regional food security in the region?

Major Issues Arising and the Reality of Co-operation amongst the Nile countries

It is clear that basin-wide co-operation is the way forward in the face of rapidly increasing population, conflicts, environmental degradation and frequent natural disasters such as drought and famine, which all exacerbate the water shortage in the Nile Basin. As Erlich states, "the very existence of dialogue on co-operation between Egypt and Ethiopia (key players in the Nile Basin), is a great achievement in itself as the actual positions of these countries, even in the NBI, will be critical"[24]. He considers such dialogue as key to the first step towards co-operation on the Nile Waters and notes, "the more Egypt and Ethiopia liberalize their views of themselves, the greater the chance for mutual understanding."[25]

Water resource disputes among member states could easily unhinge future developments. Moreover, while donor funding for the NBI is great, and is likely to increase in the future, member countries have to start thinking on how to maintain the initiative in cases where donor funding is reduced or not forthcoming.

Existence of Effective and Efficient Institutions

Egypt's willingness to join the NBI might immensely reduce its power. As a member of NBI, it seems unlikely that Egypt will maintain the dominant position it has had for the last 30 years, or any other Nile country to occupy such a position. Therefore, if co-operation is to take place amongst these countries, then it has to be co-operation on an equal basis. However, my assumption is that, as the power of a country like Egypt in making decisions reduces, the commitment of the other countries towards the success of the NBI may increase. This would, in turn, lead to a more favourable institutional environment for co-operation.

Adequate and Reliable Information

Through the Nile Basin Initiative, information exchange among member countries has improved. Agreements that seemed impossible to make before due to tensions are becoming feasible because suspicions among these countries are reduced. In the past, the upper countries did not possess the kind of information Egypt and the Sudan possessed over the Nile River.

More Members Joining the NBI

This is the first time a common vision has been established among the ten Nile basin countries. However, the challenge is how to bring member countries to dialogue. It is evident that the NBI members had foreseen the problem of many members arising in the future. That could have been the reason why they came up with Subsidiary Action Programmes, so that, in cases where these projects could not be implemented at international level, then they would at regional or national level. Despite the fact that the projects are still being implemented under the umbrella of the NBI, the numbers are smaller, thus decision- making becomes easier, and each state is given more incentives to make and keep commitments. Additionally, more attention is focused on the development of projects that will enhance the economic and social welfare of the citizens of these countries. What is important is that the NBI, in its operations, takes into considerations priority areas of individual countries.

In specific cases, it may not be necessary for all the riparian countries to reach agreements on projects within a sub-basin. In such cases, a sub - basin agreement may be appropriate with only those countries affected, including relevant upstream and downstream countries.

Political Willingness among the Nile Countries

The Nile basin countries are politically willing to co-operate because of the common challenges facing them. They are willing to carry out agreements outweighing the constraints of existing rules amongst them. This could be due to the realisation that, to attain their objectives, they have to be committed to their common vision to make co-operation possible among them. Without a doubt, this would, to a larger point, restrict individual countries' own pursuit of advantage on specific issues in the future.

Expectations of Future Benefits

Co-operation among the Nile countries is taking shape due to expectations of future benefits which could be both internal and external. Without a doubt, management of Nile conflicts, both present and future, require effective co-operation among increasingly diverse backgrounds of different countries.

Equal Distribution of Duties and Responsibilities

Duties and responsibilities among members of the NBI are being distributed equally. This has given each individual country the feeling of ownership, and this is one fact that can highly enhance co-operation and reduce suspicion among them. We see a reduction in the transaction costs of undertaking different activities within the organisation. For example, whenever there is any meeting, the hosting country will cater for all the related costs. I strongly believe that if this continues, solutions to different collective problems will be found. The feeling that each of the states has an equal representation and yields equal power within the NBI, is a positive indicator of co-operation. Even countries that had no say over the Nile River can now exercise authority and are part of decision-making.

National Interests of Individual Countries

Over the years, national interests, perceptions, and positions of different Nile countries have in the past negatively impacted on any co-operation initiatives. However, with the establishment of the NBI, Ethiopia and Egypt, who are key players, have joined in the initiative to work towards common ground. The implications being that Egypt might be forced to reduce its share of the Nile waters, and Ethiopia to stop some of the potential activities along the River, which have led to tensions between the countries.

Construction of Dams

As one of its major objectives, the NBI is also expected to heavily rely on construction of large-scale irrigation schemes to promote economic co-operation. Before this is done, NBI member states need to analyse the impact on water flows of such projects. More research should be carried out on the eventual impact on Nile water flows if a whole series of new Hydro Electric Power (HEP) plants are built. Such matters need to be considered if a long-term strategy is to be implemented. The main argument of this study is that the reduction of the Nile waters, in any way, has been a major cause of disagreement, and would remain so for many years; if much is not done to preserve the Nile waters.

Participation of Non-Governmental Organisations (NGOs) and Civil Societies

Participation of NGOs and Civil Societies in the NBI is minimal. The question is, what should the relationship between the two be like? The study has further demonstrated that NGOs greatly influence the manner in which the international system addresses environmental issues and this would be the same, within the NBI, if they became involved.

While NGOs are gaining power and recognition in the international arena, many have complained that the NBI remains an organisation of states, with only a limited role for civil society. The reasons are not clear, although the assumption is that, maybe, the Nile Basin states fear expanding the role of civil society as it may undermine their own sovereignty. Nonetheless, this is one area, which cannot be ignored by the NBI. The challenge, therefore, lies in devising an acceptable arrangement that will allow the formal participation of civil society in NBI without threatening state sovereignty.

The Financial Question

It is my view, the NBI seems to depend too much on grants and external funding particularly from donor agencies. Their major source of funds is the World Bank. Other commitments and changing priorities of the Bank might stretch the availability of funds. It is true that the NBI countries are already contributing towards some of their activities. However, the Nile countries need to do more. They need to formulate alternative financial arrangements for major sustainable capital investments. The NBI has also correctly identified an holistic approach as a crucial conflict management mechanism. However, this cannot be achieved through over reliance on foreign assistance. In order to ensure its sustainability, Community Based Organisations (CBOs) and other mechanisms should be identified and strengthened at national level. This will ensure that they not only participate in decision-making, but also ensure that their livelihood systems are sustained. This holds the promise of being cost - effective in the overall management of conflicts in the region.

Co-operation and Co-ordination at Regional Level

So far, at national level, successes of the NBI are limited, which has implication on the management of the Nile River. This is mainly attributed to uncoordinated sectoral approaches to environmental management. There exists, to a larger extent, weaknesses in the legal and institutional framework, and the lack of incentives and enforcement mechanisms among others. Therefore, there is a need to establish comprehensive legal and institutional mechanisms. In a fragmented way, individual countries of the Nile Basin have taken steps to establish such mechanisms. What is lacking is the strong support for these national initiatives. They should also be made to run parallel to the sub-regional and international initiatives.

Conclusion and Policy Recommendations

Education of the Population and Policy Makers about the Operations of the NBI

There is a variety of ways in which the NBI can enhance its contributions in the field of accountability. The NBI should, however, continue to promote such activities, because a well-informed and well-educated population combined with the presence of policy-makers/experts is the first step towards managing some of the major conflicts affecting the region.

Members' Commitment to NBI Objectives

An institution like the NBI is only as strong as its member states. Therefore, it is necessary to encourage governments of member countries to be more politically willing and committed to achieving objectives set by the NBI. This can only be done if each country within the NBI provides adequate and reliable information about the projects being implemented. As stressed, "small injections of new knowledge can play an important role in arriving at positive resolutions."[26]

Capacity Building and Use of Appropriate Technology Within Member States

Many of the NBI member states lack the capacity to address their conflicts adequately. I suggest that specific areas where expertise is needed include planning and the reconciliation of local, regional, and national priorities and capabilities.

Integrated, International Approach

An integrated approach that connects these conflicts to the more urgent concerns like environment, development and security is critical to its effective management. There should be inclusion of conflict management into the agenda of existing multipurpose regional institutions. Since these institutions are concerned with the management of political, security, and economic affairs, involving them would help in developing conflict management policies in the context of the broader regional institutional and policy frameworks. Managing the Nile conflicts demands high levels of co-operation and policy co-ordination at all levels, from the village, community, state, regional and global levels. The problem is that the necessary vertical and horizontal co-operation is often lacking.

Equitable Distribution of Water Amongst the Countries

Water should be distributed equitably and not equally to avoid a situation where certain countries have more than they need because of their geographical location. An important challenge is to weigh domestic, social, political, and economic factors in the distribution and management process.

Individual Country's Priority Areas

Many NBI activities may be difficult to implement collectively given the varied backgrounds and interests of the members. Within the NBI, a certain amount of system review still needs to be done to determine priorities and ensure that they are adequately addressed before other areas that might not be of priority. Among the members, some support strengthening of existing institutions, while others prefer the establishment of new ones. There are calls to expand the NBI's work in new areas, such as energy, fresh water resources and sustainable tourism. Some experts call for convening more conferences and negotiating more treaties. Personally, I feel that the first issue that needs redress is modification of the outdated Nile treaties. Notwithstanding its limitations, the NBI can be and has been a critical force in raising national and regional awareness of the conflicts in the Nile region and in urging the need for and benefits of co-operation in addressing them. Its role is more significant than ever, and if successful, has the potential to become an important instrument for resolving some of these conflicts, which have always afflicted the Nile region.

In conclusion, unless national interests of individual countries are taken into consideration, the success of their co-operation will remain a myth. The existing information exchanges and sharing of ideas is not adequate and should be upgraded. The commitment to one another and to the mission of the NBI can only be sustained through political willingness, mutual reciprocity and collaboration. Equal distribution of duties and responsibilities is a good trend that should be maintained.

Notes

1. Smith, M., J. 1996. *Nine Nations, One Nile.* p. 3

2. Available at http://www.umich.edu/~csfound/545/1996/smith.html (1 June 2006).

3. Available at http://www.umich.edu/~csfound/545/1996/smith.html (1 June 2006).

4. Dahilon, Yassin, Mohamoda, 2003. *Nile Basin Co-operation: A Review of the Literature.* Nordic African Institute, Uppsala, p. 7.

5. Smith, M., J.

6. Ashok, Swain. 2004. The Nile River Basin Initiative: Too Many Cooks, Too Little Broth. *SAIS Review, Vol. XXII,* No. 2, pp. 93-308.

7. The Nile River Basin Initiative: Too Many Cooks, Too Little Broth. *SAIS Review, Vol. XXII.*

8. Erlich, Haggai. 2002. *The Cross and the River: Egypt, Ethiopia and the Nile.* Boulder, p. 8.

9. Dahilon, Yassin, Mohamoda, 2003. *Nile Basin Co-operation: A Review of the Literature.* Nordic African Institute, Uppsala, p. 7.

10. Dahilon, Yassin, Mohamoda.p. 26.

11. Dahilon, Yassin, Mohamoda.

12. Waterbury, John. 1990. *Legal and Institutional Arrangements for Managing Water Resources in the Nile Basin.* In The Middle East and North Africa: Essays in honor of J.C Hurewitz. New York: Middle East Institute, Columbia University. pp. 276-303.

13. Waterbury, John. 1990. *Legal and Institutional Arrangements for Managing Water Resources in the Nile Basin.*

14. Collins, R.O. 1990. *The Waters of the Nile: Hydropolitics of the Jonglei Canal, 1900 - 1988.* Oxford: Clarendon Press.

15. *The Daily Nation,* September 7, 2002; The Monitor, September 7, 2002.

16. Dahilon, Dahilon, Yassin, Mohamoda. p. 27.

17. Dahilon, Dahilon, Yassin, Mohamoda.

18. Tafesse, Tesfaye. 2001. *The Hydropolitics Perspective of the Nile Question*. p. 67.

19. Biswas, Asit. K. 1992. Water for the Third World Development. *Water Resources Development, Vol.* 8, pp. 6, 34-36.

20. Tafesse, Tesfaye. *The Hydropolitics Perspective of the Nile Question*. http://chora.virtualave.net/tafesse-nile.html (Accessed, 29 April, 2009).

21. Hultin, J.1995. The Nile: Source of Life, Source of Conflict. In *Hydropolitics: Conflict over Water as a Development Constraint*. Dhaka University Press Ltd: Dhaka, pp. 2 9-54.

22. Dahilon, Dahilon, Yassin, Mohamoda. pp.12 -14.

23. Collins, R.O. p. 246.

24. Erlich, Haggai. p. 218.

25. Erlich, Haggai

26. Chasek, Pamela. 2000. *Global Environmental Politics*. Boulder: Westview Press.

Managing Trans-Boundary Water Conflicts on Lake Victoria with Reference to Kenya, Uganda and Tanzania

Kakeeto A. Richard, Chichaya Hope, Nikodemo Andrea & Ndungu Catherine Njeri

Introduction

Lake Victoria is a shared water resource and confrontation on any of its islands or beaches has the potential to manifest itself as an internationalized conflict in the region. The cross-border conflicts on Lake Victoria are part of a conflict system whose epicenter has to be diligently identified if the said conflicts are to be well - managed.

This article is an attempt to locate the conflict over fish resources as the epicenter of the cross -border conflicts in the Lake Victoria region. The article further argues that other conflicts whether diplomatic, regional or environmental only serve to fuel, aggravate, or mitigate the epicenter. Any attempts at management of cross - border conflicts on Lake Victoria should accordingly have as their target, the sustainable and equitable use of its resources. The article on one hand acknowledges what has already been done in the direction of cooperation and on the other, argues that the principles underlying the formation of these cooperative efforts should be more pronounced at all possible opportunities. For the purpose of this article, we will use the definition of conflict as a confrontation between one or more parties aspiring towards incompatible or competitive means or ends.[1] The article further points out the qualities of Lake Victoria as a shared resource then explores the real conflicts over the Lake with special focus on the Kenya- Uganda dispute over Migingo Island. In addition, the article will mention the water hyacinth as one of the indicators of environmental conflict and then highlight the potential conflicts over the lake and the anomalous structures in which they persist. The article finally investigates the causes of conflict on Lake Victoria, attempts at cooperation and how they can be strengthened. To the shared resource we now turn.

Background Information on Lake Victoria as a Shared Resource

Lake Victoria is the largest tropical Lake in the World and second largest fresh water lake by surface area. It is 68,800 sq km and is shared by Tanzania (51%), Uganda (43%) and Kenya (6%).[2] The waters of the lake come from 82% precipitation and 18% surface runoff. In respect to surface runoff, rivers play an important role and of these, sixteen are from Kenya, ten from Tanzania and seven from Uganda. The Kagera River is another major contributor to the waters of the Lake as it contributes 7% of the waters of the lake.[3]

The waters of Lake Victoria are also shared by the lower riparians of the River Nile which starts off its journey to the Mediterranean from Jinja in Uganda. The

Colonial regime governing the relationship of the upper and lower riparian states on the Nile has always been a source of disputes. The upper riparian states have often argued that the colonial agreements do not bind them while the lower riparians argue that the treaties are binding as a result of succession to the treaties of the former colonies upon independence. All the Lake Victoria States issued formal repudiations of these colonial treaties as typified by the reaction of Uganda, immediately after independence.[4] For this reason, the dynamics of conflict on Lake Victoria are better captured when placed within the larger Nile basin as shall be expounded later.

The name given to this second largest lake in the world signifies a shared history of the peoples living around it. The Lake had several names given to it by people from the three countries: Lake *Lolwe* (by Kenyans), Lake *Nalubaale* (by Ugandans), *Sukuma Lake* (by Tanzanians) and Lake *Ukerewe* (Arab name). The first Europeans to reach the source of the River Nile in 1858 were British explorers lead by John Hanning Speke, who was part of an expedition team searching for the source of the Nile, an important strategic knowledge for the British Colonial administration at the time. John Hanning Speke decided to name the Lake "Victoria" after the Queen of the England. The name Lake Victoria was popularized in the Western World by an American journalist, Henry Morton Stanley, and the name has endured to date.[5]

Lake Victoria catchment area is approximately 38,000 km^2. It has a varied topology ranging from valleys, plains, hills, plateaus to mountains. The general gradient varies from a slope of approximately 0% to 30%. More than 78% of the total land area in the region lies within the mid slopes (2% - 18%) while the rest of the 22% lies within the critical brackets of 0 – 2% and 18 – 30%. This represents the single most important catchment area for the Lake because it provides nearly 75% of the total water volume flowing into the Lake through the River systems.[6]

By 2005, the Lake Victoria basin was estimated to be supporting 35 million people.[7] The economic opportunities within the Lake Victoria region include: agriculture, aquaculture, water and sanitation, tourism, industrial development, trade and commerce, mining, research and education. The activity that stands out in relation to conflicts is that of fishing and fisheries resources. We now consider the real conflicts.

Real Conflicts

Real conflicts are those distinguishable from mere competition since the latter only degenerates into conflict when parties try to enhance their own position by reducing that of others or by putting them out of business. This is a key tool in evaluating the conflicts on Lake Victoria. Conflict is also distinguishable from tension. The latter always precedes and accompanies the former but they are never the same and are not incompatible with cooperation.[8]

The research that has been done by the World Conservation Union and Lake Victoria Fisheries Organisation indicates a wide range of conflicts all of which are related to the fishing resource.[9] The point of convergence of all the conflicts seems to be the fishing industry. It ranges from fishermen-fishermen conflicts and fishermen-fish agency conflicts to interstate tensions. Some of the alleged conflicts on the Lake start out as mere arrests for violation of the law but end up depicting a situation of conflict. This occurs in situations where the arrested individuals are nationals of another partner state on Lake Victoria.

Other reported conflicts include those over fishing gear and piracy. As long as, among the aggressors and victims, there are individuals from across the borders, this is treated as a cross-border conflict. At Maduwa on the Kenya-Uganda border, it was reported that Kenyan fishersmen would steal nets of their Ugandan counterparts and also use illegal gear in fishing.[10] The Kenyan fishermen would also complain that their Ugandan neighbours at Marenga use the Uganda Special Revenue Protection Services (SRPS) to confiscate their fish which is sold off cheaply to the Ugandan traders.[11]

Kenyan fishermen have found it very difficult to fish in the Ugandan waters due to intensified patrolling by the Ugandan government since 2001. It is however, the case that arrests and confiscation of fishing gear affects both the fishermen on the Kenyan and Ugandan sides. Fishermen on both sides have indicated that government agencies like the SRPS of Uganda extort money from them during their surveillance operations.[12] These ping-pong accusations go on wherever there are islands in areas with unclear boundaries.

Similar concerns are expressed at the Kenya -Tanzania Border. Fishermen from either side of the border at Mugabo Island do not see themselves in conflict. The Kenyan fishermen however, face what they perceive as conflict with the Tanzanian authorities in respect to access to the Tanzanian waters for fishing.[13] At Sota Island on the same border, fishermen are concerned about how drift net fishing of Kenyan fishers affects the nets of Tanzanian fishermen.

At the Tanzania-Uganda border, conflicts are mainly related to theft and destruction of fishing gear. The different types of nets used for fishing on either side of the border may account for dispute. There have been more reports of arrests of Ugandan fishermen by Tanzanian authorities than those by Ugandan authorities. Piracy is another concern on the Tanzania-Uganda waters.[14]

In the 2004 Strategic Conflict Analysis, the conflicts above were referred to as smaller skirmishes in defense of own fishing rights, the value of which, even though involving some shooting, could not be sufficient for justifying a full - scale interstate war.[15] It is however, the argument of this article that had cooperative mechanisms within the region not existed; "a skirmish" at the Migingo Island would have escalated into a protracted conflict.

The Migingo Question

To the fishermen, little importance is put on the international boundaries. They are least bothered by where the boundaries are until they are arrested on foreign waters. The dispute over the island ownership has led to the formation of a joint survey task force to review border demarcation. From appearances, one standing at Sori Bay in Migori District would see the island which is located approximately 10km away.

However, appearances and occupancy may not be enough to settle such a matter. The International Court of Justice (ICJ, the Hegue, Netherlands) was faced with a similar situation in the Nigeria-Cameroon dispute over the Bakassi peninsula. The Nigerian community had confused occupancy and proximity with ownership. In the end, the ICJ declared that the Island actually belonged to Cameroon.[16] In such situations of dispute, Uganda's deployment of security forces on the island was destined to result into tensions. Actually the trigger was when 400 Kenyan fishermen were evicted by Ugandan authorities for failure to pay fishing fees.[17] State Minister for Internal Affairs in Uganda, Matia Kasaija, stated: *"Kenyans cannot claim this Island because it belongs to Uganda until the colonialists decide otherwise. In fact, we have instructed the Lands Minister to sort out the problem and we shall invite the Kenyans to a joint re-surveying exercise of the island to determine the owner."[18]*

Uganda is thus not treating this matter as a foreign affairs issue per se, it has major territorial and ultimately security implications. A committee of the East African Legislative Assembly met in Rwanda in February 2009 to evaluate reports by surveyors and then give recommendations. Article 5 of the Ugandan Constitution[19] and a schedule there too describes the boundary yet the Kenyan constitution seems to be silent. Ugandan authorities say the island falls within the boundaries of the district of Bugiri. Indeed the Google earth maps make the matter more complicated. One report cites the Google earth map as placing Migingo islands within Ugandan boundaries at coordinates 2°48'06.82"S and 32°38'45.25"E.[20] Another reference to the Google earth map places Migingo Island on the Kenyan side of the border at coordinates 0°54'26.40"S and 33°56'46.22"E.[21] It then seems safe not to draw any conclusions yet. It is however, imperative to wait for a joint survey report since this state of affairs requires diplomatic diligence if it is not to yield into a situation of conflict. Yet, not all conflicts are as manifest as the one over Migingo Island. Other conflicts are subtle and involve not human beings but the environment.

Environmental Conflicts: The Water Hyacinth

If the epicenter of conflicts on Lake Victoria is the fisheries resource, then it must be the case that any state of affairs human or natural that affects this resource, would occasion conflict. The water hyacinth has occasioned an antagonistic state of relationships between individuals, members of society or governments in connection to the utilization of Lake Victoria as a shared resource. Its characteristic spread across the lake limits the fishing grounds and hence environmental conflict arises from natural resource constraints. Its other characteristic is the ability to suffocate other species with the effect of causing environmental conflict through environmental degradation.

Margaret Oduk, a research scientist at the United Nations Environment Programme (UNEP), in describing the impact of the water hyacinth, states that the continued invasion of the weed might result in food insecurity as it blocks access to fishing ground and interferes with water transport system due to its rapid growth in the Lake Victoria.[22] Mr Obiero Onganga, the Executive Director of OSIENALA (Friends of Lake Victoria), a local environmental NGO based in the lakeside city of Kisumu, described the weed as causing heavy ecological havoc in the Lake.

The rapid growth of the weed produces long roots and tissues in its stem which enable it to float in water freely. It grows in clusters that form floating mats in the Lake and thrives best in polluted waters. The water hyacinth not only threatens the Lake but is also an ecological disaster that overwhelms the Lake and its biodiversity.

The origins of the water hyancith on Lake Victoria are not very certain. It is a recurrent, free, floating water plant first seen in South America. The plant is one of the threats facing Lake Victoria and its environs. It grows with other aquatic weeds and has a high capability of rapid growth. The water hyacinth is a large, fleshy, fresh water plant capable of dynamic growth. In Africa, the water hyacinth has been present in significant quantities for several decades, mostly in major rivers and in fresh coastal waterways. It first appeared in Egypt's Nile in 1890's where initial introductions were deliberate as it was used for decorative purposes. In 1908, it was seen in Natal province, South Africa and in the 1930's it appeared in several lakes in Zimbabwe. In 1950's the hyacinth colonized the Rivers Congo, White Nile in Sudan and Pangani in Tanzania.

In East Africa, the water hyacinth was sighted in 1982 in Lake Naivasha and Lake Kyoga, Uganda. In Tanzania it appeared in 1990 and in Rwanda in 1987. It has since multiplied rapidly to cover several locations of Lakes Victoria, Albert, Kyoga and the shores of River Nile.

Potential Conflicts

Potential conflicts, mainly an act of postulation, though unlikely in the immediate future are worth thinking about. They include "water wars" which have been posited by reputable persons like Boutros Boutros Ghali, especially in respect to the Nile basin.[23] Increased demand in the upper riparian of the Nile, it is argued, may challenge Egypt's interests and provoke violent reactions. Of course the contribution of the Victoria basin to the consumption of Egypt is lower than that of other tributaries. But, it is true that increased water conflicts on the lower Nile between Egypt and Sudan, Egypt and Ethiopia, may have the effect of spilling over into the East African region that are partner states on Lake Victoria. If such spill over has the effect of destabilization then such potential conflicts are not as remote.

Within the conflict system[24] in the Lake Victoria region, a conflict over land for instance in the Mau forest of Kenya and environmental degradation there has a direct impact on the amount and quality of flow into the Lake. As shall be shown in the causes, irrigation, and the amount of water released into the Lake and ultimately into the Nile may all may be indicators for future conflicts. International law favors equitable utilization of international water courses as opposed to both absolute territorial sovereignty and absolute territorial integrity.[25]

Egypt however, in respect to the Nile whose hydropolitics is inseparable from Lake Victoria, seems to argue that the 1929 Nile waters agreement created servitude in favour of Egypt as against the territories of the upper riparian states.[26] This anomalous state of affairs could, without a full blown war, impede efforts towards cooperation and hold conflict in potency in the region. It would then be important to investigate the causes of the said conflicts.

Causes of Conflicts

Fish and fisheries resources stand out as the epicenter of the conflicts over Lake Victoria. The research so far carried out indicates that there is a correlation between the increased trade in the Nile Perch and the conflict on the Lake.[27] It advanced that before the steep rise in trade of the Nile perch, fishers seemed to interact and cross borders with comfort and less confrontation.[28] Ogutu observes that overall fish catches increased fivefold between 1975 and 1990, making Lake Victoria the single most important source of freshwater fish in the world, with the Nile perch as the leading export.[29]

Increased fish harvesting and corresponding demand for fresh water fish led to the establishment of processing plants for Nile perch, mainly for export, and stimulated fishing. Ironically, this rapid increase in fishing effort is now not only a major threat to the perch fishery, but the competition over this diminishing resource is causing tensions among the fishing communities.

As earlier indicated, tensions in fisher communities at the borders cannot be taken lightly given their potential to escalate into internationalized conflict. To these fishing communities, biodiversity is not a question, to them, the pricing, availability and reliability of the lucrative Nile perch seems to determine all they do in the fisheries.

Actually the tensions on Migingo Island earlier discussed have an economic bearing. It seems there have been more fish landings for export on the Kenyan side than on the Ugandan side. Even though the principle is that fish should land where it is fished, the lucrative Nile perch has been fetching better prices on the Kenyan side so that even Ugandan fishers have been landing on the Kenyan side to sell at a good price. The implication is obviously loss of income from licenses and fees by the Ugandan government.[30]

The failure to balance the establishment of fish processing plants and the fishing mechanisms on the lake is part of the causes of conflict. If some of the larger fishers in Kenya use the trawler method of fishing yet other partner states do not, the effect will be a reduced catch in its 6% portion of the lake. The artisanal fishers in Kenya will then have to find fishing grounds in areas outside the border. Even though the cross-border interactions through beach management units have proposed that fish should land where it is fished, this policy may not necessarily apply where the competition for fish is between larger fishers and the artisan fishers.[31]

Unclear boundaries, ignorance of fisheries laws regulating access to fishing grounds coupled with poor communication between authorities on both sides of the border are a recipe for the conflicts on the lake. The propensity for conflicts is also heightened by income disparities within the fishing communities on either side of the border, which were sighted as one of the reason for the conflicts.[32] The ramification of this precondition for conflict is in the resultant smuggling, piracy and destruction of each other's gear by fishers.

A drop in the levels of the Lake reduce the fishing ground available to fishers in all the partner states.[33] This increases contiguity of fishermen from the partner states and conflict is inevitable. This cause of conflict on the lake is deeply entrenched in the man - made activities within the basin. Deforestation and poor farming methods within the basin reduce the amount of water available to recharge the lake. Recent forest fires in the Mau forest have an impact on the ultimate amount and quality of water that gets to Lake Victoria.[34]

Conflicting interests over the lake among the partner states may account for the drop in the levels. If it is the case that Uganda has increased the amount of flow to Egypt, then, this coupled with irrigation in Shinyanga, Tanzania, the level of water has to fall. The fall in the level results into the shortages and ultimately conflict. Uganda for instance seems to be bound under the 1949 and 1953 Owen Falls Agreements which turned Lake Victoria into a water reservoir for Egypt provided the latter participated in the construction of the dam. Waters in a reservoir can be released at the wish of the party for whom such waters were reserved which may occasion conflict of interest. Despite all of this contest and competition over the water resources, some efforts have been made towards cooperation.

What has been done?

In the last three decades, some activities have taken place regarding the Lake Victoria Basin. Three important international agreements have been concluded: the Kagera Basin Organization (KBO) was established in August 1977 through the Rusumo Agreement signed by Rwanda, Burundi and Tanzania with Uganda joining in 1981. In practice, the Organisation has so far focused on developing of large water projects, including the developing of hydropower at Rusumo and irrigation projects. According to Okidi,[35] the KBO Agreement did not establish an elaborate arrangement for fundraising to finance common works as is in the case of other bodies. Each member was to contribute an equal proportion of the administrative budget. Contributions from member states were not forthcoming yet the KBO would depend, in the main, on donor grants to implement its projects.

The Convention for the Establishment of the Lake Victoria Fisheries Organisation, which was adopted by partner states at Kisumu, on 30 June 1994, established the Lake Victoria Fisheries Organisation (LVFO) for two reasons: first, to harmonise fisheries policies and legislation and second, to promote the conservation of the lake environment in general and set the standards to be achieved by the three member states including most of the general principles of international water law.

The LVFO has as its main principal the mission to foster cooperation among the Contracting Parties, harmonize national measures for sustainable utilization of the marine resources on the Lake and to develop and adopt conservation and management measures.

The Agreement on the Preparation of a Tripartite Environmental Management Programme for Lake Victoria of 5 August, 1994, brought into existence Lake Victoria Environmental Management Project (LVEMP), which is due to start its second phase. This latter phase is expected to cover a longer period of implementation, drawing lessons from and consolidating the successes gained during its first phase. It is this project that attempted to facilitate activities designed to control the water hyacinth. The control of this weed included the Biological method, the chemical method and the manual method. Each of these methods had its own shortcomings but the ultimate limitation to the control of the hyacinth was uncoordinated responses from the partner states.

Most important is the Protocol for Sustainable Development of Lake Victoria Basin signed by East Africa Community (EAC) partner states on 29 November, 2003. The protocol, under the auspices of the East African Community treaty, governs the partner states' cooperation in sustaining the Lake Victoria Basin. Yet, despite these agreements and forum of cooperation, the integrated management of the lake has not been streamlined since tension and strife still exist among partner states on Lake Victoria. We now propose what ought to be done.

What ought to be done?

A lot is on paper in relation to the co-operative effort of managing Lake Victoria conflicts. However, partner states must be convinced about the principles upon which these regional documents have been drawn. These principles are now reflected in the United Nations Convention on Non-Navigational Uses of International Water Courses of 1997. The convention is applicable to most watercourses and a diligent observance by partner states would be fruitful in avoiding conflict.

If the *principle of international co-operation* by watercourse of international water resources is taken seriously, then conflicts would be minimised. The Convention on Non-Navigational Uses of Water Courses provides that "Watercourse States shall cooperate on the basis of sovereign equality, territorial integrity, mutual benefit and good faith in order to attain optimal utilization and adequate protection of an international watercourse".[36] The conceptualization of the *concept of an international watercourse,* under the Stockholm Declaration has to be explained to the stakeholders and in particular to the politicians. They need to know that it is not only surface waters that matter but also ground waters provided they flow into a common terminus.[37] The existing co-operative efforts would then seek to challenge any promotion of deforestation and poor farming methods in any of the states.

The partner states should be reminded of their *obligation not to cause significant harm to co-riparians.* In utilising resources, states are required not to cause significant harm to the interest of other states by pollution or other conduct[3.8].Neither can it be overstated that partner states have a duty to *protect all reasonable and beneficial uses of the waters.*[39] While wasteful uses of water are discouraged, in the event of a conflict between uses of an international watercourse, special regard is given to the requirements of vital human needs.[40]

Partner states have been guilty of failure to share information on upcoming activities for which the principle of *notification and information sharing,* would be advisory to the them.[41] This right of a potentially affected state to demand notification in order to safeguard its interests was restated by the Arbitral Tribunal in the *Lake Lannoux case.*[42]

Conclusion

In this paper, we have explored the conflict over fisheries resources as the epicenter of cross-border conflicts in the Lake Victoria region. The paper has highlighted the real and potential conflicts pointing out the Migingo question as a demonstration of what mere skirmishes at a border island could portend for the whole region. It has also indicated the effect of the water hyacinth on the conflict system and stressed that whatever happens within the territorial jurisdiction of any of the partner states has a direct impact on the Lake and ultimately on the fisheries resource. It has been found out that a lot of initiatives for cooperative management of the lake basin are in place but that the principles underlying such initiatives should be reiterated over and over again for decision makers to be held accountable.

Notes

1.   Miller, Christopher, E., & King, Mary, E. 2005. *A Glossary of Terms and Concepts in Peace and Conflict Studies.* UPEACE Africa Pogramme: Addis Ababa. p. 22.

2.   Lake Victoria Basin Commission (LVBC). 2007. *Shared Vision and Strategy Framework for Management and Development of Lake Victoria Basin.* Popular Version. p.1

3.   Lake Victoria Basin Commission (LVBC). 2007.

4.   Godana ,B., A. 1985.   *Africa's Shared Water Resources.* London: Frances Pinter Publisher. pp 147-156.

5.   Oyugi, Aseto & Obiero, Ong'a ng'a. 2003. *Lake   Victoria   and   its   Environs: Resources, Opportunities and Challnges.*Kisumu: OSIENALA (Friends of Lake Victoria). p.12.

6.   Obiero, Ong'ang'a & Othieno, Fredrick &   Munyirwa, Kinya (eds.). 2001. "Lake Victoria 2000 and beyond. Kisumu: OSIENALA (Friends of Lake Victoria). p.10.

7.   Lake Victoria Basin Commission (LVBC). 2007.

8.   Dougherty, James E., & Pfaltzgraff, Robert L., Jr. 2001. *Contending Theories of International Relations. A Comprehensive Survey.* NewYork: Longman. pp. 189-190.

9.   Heck, S., et al. Cross-border Fishing and Fish Trade on Lake Victoria. *IUCN/LVFO Fisheries Management Series Vol. 1,* July, 2004. pp. 28-63.

10. *IUCN/LVFO Fisheries Management Series Vol. 1,* July, 2004. p. 29

11. *IUCN/LVFO Fisheries Management Series Vol. 1,* July, 2004.

12. *IUCN/LVFO Fisheries Management Series Vol. 1,* July, 2004.

13. *IUCN/LVFO Fisheries Management Series Vol. 1,* July, 2004. p. 31.

14. *IUCN/LVFO Fisheries Management Series Vol. 1,* July, 2004. pp. 37-38

15. Swedish International Development Agency (SIDA). 2004. *Lake Victoria Region: Strategic Conflict Analysis.* Department for Africa. p.77.

16. Bouknegt, Thijs. "Nigeria to hand over Bakassi to Cameroon": Available at http://www.rnw.nl/internationaljustice/courts/ICJ/080813-nigeria-cameroon (Accessed March 28, 2009).

17. Mubatsi, Asinja ,Habati. " Politics of Fish in Migingo Island Dispute" *The Independent,* Kampala 10 March, 2009: Available at http://www.independent.co.ug/index.php/news/news-analysis/79-news-analysis/688-politics-of-fish-in-migingo-island-dispute (Accessed on 28 March, 2009).

18. Yasiin, Mugerwa. "East Africa: Uganda, Kenya Fight for Migingo Island". *The Monitor,* October 21, 2008.

19. Second Schedule of the Constitution of The Republic of Uganda as amended, 1995.

20. Mubatsi, Asinja ,Habati. " Politics of Fish in Migingo Island Dispute" *The Independent,* Kampala 10 March, 2009: Available at http://www.independent.co.ug/index.php/news/news-analysis/79-news-analysis/688-politics-of-fish-in-migingo-island-dispute (Accessed on 28 March, 2009).

21. Available at http://images.google.co.ke/images?hl=en&q=migingo+island&btnG=Search+Images&gbv=2 (Accessed on March 31, 2009).

22. Oyugi, Aseto & Obiero, Ong'a ng'a. 2003. pp. 110-111.

23. Mike, Thompson. "Ex-UN Chief warns of Water Wars". BBC News, 2 February, 2005: Available at http://news.bbc.co.uk/1/hi/world/africa/4227869.stm (Accessed on 26 March, 2005).

24. The conflict system perspective of conflict analysis acknowledges that every conflict is interconnected with other conflicts within the region such that management of such conflict must trace the larger regional pattern of a conflict. Makumi Mwagiru. 2000. *Conflict, Theory, Processes and Institutions of Management.* Nairobi: Watermark Publications. p. 72.

25. Ntambirweki, John. (1997 Unpublished). *The Emerging Legal Framework on the Nile: Towards an Integrated River Basin Management Regime.*

26. Dr Fahmi, Aziza. 1967. "International River Law for Non-Navigable Rivers with Special reference to the Nile". *Revue Egyptienne de Droit Internationale, Vol. 23,* 39. p. 53.

27. Heck, S., et al. Cross-border Fishing and Fish Trade on Lake Victoria. *IUCN/LVFO Fisheries Management Series Vol. 1,* July, 2004. p.29.

28. Heck, S., et al. Cross-border Fishing and Fish Trade on Lake Victoria. *IUCN/LVFO Fisheries Management Series Vol. 1,* July, 2004.

29. Ogutu-Ohwayo, Richard. *The Fisheries of Lake Victoria: Harvesting Biomass at the expense of Bio-diversity:* Available at http://www.unep.org/bpsp/Fisheries/Fisheries%20Case%20Study%20Summaries/Ogutu (Summary).pdf. (Accessed on 26 March, 2009).

30. Heck, S., et al. Cross-border Fishing and Fish Trade on Lake Victoria. *IUCN/LVFO Fisheries Management Series Vol. 1,* July, 2004. pp. 28-63.

31. Heck, S., et al. Cross-border Fishing and Fish Trade on Lake Victoria. *IUCN/LVFO Fisheries Management Series Vol. 1,* July, 2004.

32. Heck, S., et al. Cross-border Fishing and Fish Trade on Lake Victoria. *IUCN/LVFO Fisheries Management Series Vol. 1,* July, 2004.

33. International Rivers Network. "Dams Draining Africa's Lake Victoria". 9 February, 2006: Available at http://www.internationalrivers.org/en/node/1056 (Accessed on March 31, 2009).

34. UNEP. "Forest Fires Destroy Kenya's Key Water Catchments". 25 March, 2009: Available at http://www.unep.org/Documents.Multilingual/Default.Print.asp?DocumentID=573&ArticleID=6109&l=en (Accessed on 27 March, 2009).

35. Okidi, C.O. *Development and the Environment in the Kagera Basin under the Rusumo*

Treaty. Institute of Development Studies, University of Nairobi. Discussion paper No. 284, September 1986.

36. UN Watercourses Convention, Article 8(1) and Article 2(4). Convention for the Establishment of the Lake Victoria Fisheries Organization, Articles 2(2).

37. Stockholm Declaration, 1972, Principle 21.

38. UN Convention on Non-Navigational Uses of International Watercourses, Article 7.

39. UN Watercourses Convention, Articles 3, 4, 5, and 6.

40. UN Watercourses Convention, Article 10.

41. UN Watercourses Convention, Article 12.

42. Lake Lanoux Arbitration (France vs Spain). Arbitral Tribunal, 16 November, 1957.

An Attempt Towards Management:
An Examination of the Existing Institutional Frameworks in the Lake Victoria Region

Daniel Peter Lesooni and Christopher Ogachi

Introduction

This paper is an attempt to describe the various institutions that manage Lake Victoria. The description of various institutions points out a common platform where various countries, particularly Eastern African countries, cooperate for the benefits of the Lake, guided by their interests. Cooperation encourages collaboration which might lead to substantive agreements and coordination among the riparian states. Despite cooperation among these states, conflict is inevitable and countries around Lake Victoria have come up with various initiatives to encourage cooperation.

This paper emphasises the importance of managing Lake Victoria in a cooperative spirit since it has shared opportunities, for all riparian states, from a social, political and economical aspect.

The Lake Victoria Environmental Management Project (LVEMP)

The LVEMP is a comprehensive programme that covers Lake Victoria and its catchments in Kenya, Tanzania and Uganda. The LVEMP is a project that deals with both regional environment and national resources to be implemented in the Lake region. The work of the LVEMP is to clean up the lake and its catchments and manage the ecosystem sustainably. This includes reducing human waste from both urban and rural areas, soil erosion, industrial effluent, eutrophication, algae levels and water hyacinth. In general the LVEMP tries to rehabilitate the lake catchments.

The LVEMP commenced its activities in 1994 with a tripartite agreement effective from 5 March, 1997 with a $70 million[1] support from the International Development Association (IDA) and Global Environmental Trust Fund (GEF). The components of the LVEMP include management, research and extension of fisheries policies laws and their enforcement, water hyacinth control, water quality and ecosystem management and land use and wetland management.[2]

The project's main objectives include the sustainable development of riparian communities through the use of resources within the basin to generate food, employment and income, supply safe water and sustain a disease-free environment. It also aims at conserving biodiversity and genetic resources for the benefit of the riparian and global community. After years of project implementation, it managed to achieve the following: hyacinth infestation has been controlled by at least 80% i.e. mechanical harvesting at Owen fall Dams, Port Bell and

Kisumu ports through the use of hyacinth weevils to control weeds.[3] However, the hyacinth still remains the major challenge facing the riparian states.

The LVEMP has initiated several micro projects in the areas of health, education, sanitation and water supply, access to roads, fisheries and aforestation. In furthering its projects, the LVEMP encouraged community participation through planting of trees around the basin area. In addition to this, it harmonised fisheries legislation for efficiency. LVEMP II is now in the process of incorporating stakeholders and the community around the lake basin into its activities.

The Organisation set-up consists of a National Secretariat, assisted by an Operations Officer, Assistant, Accounts Assistant and a Secretary. The Support staff includes three drivers and an office assistant.[4] Its mission is to achieve environmentally and socially sustainable economic development for Lake Victoria Basin (LVB) so as to restore a healthy, varied ecosystem that is inherently stable and can support sustainable human activities in the catchments and in the lake itself. The LVEMP receives funds from the International Development Association (IDA), Global Environment Facility (GEF), the World Bank and the governments Tanzania, Kenya and Uganda. Other agencies and donors are like the Swedish International Development Agency (SIDA).

The LVEMP faces certain challenges and limitations beginning with harsh environmental conditions in the catchment area. This is due to the pressure from the increasing population which has led to severe degradation of resources as seen in the human waste which is increasingly being drained into the Lake and agricultural activities that are encroaching on the wet lands that serve as catchment areas for the Lake. Poverty and lack of awareness amongst the people living around the lake is an obstacle to sustainable development of the area. The LVEMP has a huge responsibility of dealing with most of the problems affecting Lake Victoria. It is faced with the challenge of reducing all adverse economic impacts on the lake, and therefore follow-up on various projects implemented is not always done. Finally, the LVEMP is faced with problems of integrating the objectives and priorities of the programmes into national and local development plans of partner states due to various, divergent interests of the three riparian countries.

The successes of the LVEMP have included the signing of a tripartite agreement for its implementation by the three East Africa countries. It managed to strengthen institutions through capacity building and involvement of stakeholders in forest resource management through formation of local community based institutions and manpower development. It has also developed a pollution control manual where it incorporated the Water Sewage Corporation Civil Engineering Department and the Institute of Environment and National Resources of Makerere University, Uganda so as to address the issue of water quality. The LVEMP has also supported Beach Management Units (BMU).[5]

The BMU are legally empowered to enable fishing communities participate in fisheries management and development at the beach level. The LVEMP adopted the required food safety procedures on fishing treatment which include washing, sorting filleting, skinning and trimming of fish both for export and local use. Finally, the LVEMP succeeded in standardising the size of fishing nets that will ensure the trapping of mature fish only. However the use of illegal mesh by the fishermen still poses a challenge to this effort.

The Lake Victoria Fisheries Organization (LVFO)

The LVFO came into existence due to the need, of partner states, to manage fish resources. The partner states entered into a Convention in 1994 following three seminars held between 1991 and 1995 under the auspices of the FAO-CIFA subcommittee on Lake Victoria. This led to the creation of the Lake Victoria Fisheries Commission and later to the establishment of the Lake Victoria Fisheries Organisation (LVFO) on 30 June, 1994 in Kisumu, Kenya.[6] The partner states that established the LVFO are Kenya, Uganda and Tanzania together with the Food and Agriculture Organisation (FAO), the European Union through the Lake Victoria Fisheries Research Project (LVFRP), the World Bank and the Global Environment Facility under the Lake Victoria Environment Management Project (LVEMP).[7]

The LVFO is an institution under the East Africa Community (EAC), with a permanent secretariat office in Jinja, Uganda, that facilitates a common resource management system, restoring and maintaining the long-term health of its ecosystem. The supreme body of the LVFO is the Council of Ministers, which convenes once every two years and is responsible for fisheries in the three East Africa partner states. Other organs of the organisation are: the Policy Steering Committee, the Executive Committee, Directors of fisheries management research and other committees such as the Fisheries Management Committee, the Scientific Committee and the National Committee for Lake Victoria Fisheries.

The objective of the LVFO is to foster cooperation among the contracting parties, harmonize national measures for the sustainable utilisation of the lake, develop and adopt conservation management measures.[8] The LVFO has the following functions and responsibilities: promotion of proper management and optimum utilisation of the fisheries and other resources of the Lake, enhance capacity building of existing institutions. The LVFO adopts and advocates for an ecosystem approach to management research and development of the fish resources. The LVFO vision seeks to ensure that there is a healthy and sustainable lake ecosystem, integrated fisheries management, co-coordinated research programs, generation, flow and exchange of information and an institutional/ stakeholder partnership.[9]

It also provides forums for discussing the impact of initiatives dealing with environment and water quality of the Lake basin; maintains a strong liaison with existing bodies and programmes and provides platforms for research. The LVFO encourages, recommends, co-ordinates and undertakes training and extension activities in all aspects of fisheries. It also advises on the effects of direct or indirect introduction of non-indigenous aquatic animals and plants into the waters of Lake Victoria.[10] The LVFO therefore, serves as a data bank for information on Lake Victoria fisheries and undertakes other functions as it may deem necessary or desirable in order to achieve its goals.[11]

The LVFO has established a number of groups, working on thematic areas, which operate at a national and regional level. Their work is to prepare status reports, standard operating procedures and guidelines that are then harmonised regionally and approved by the Council of Ministers for field use with regard to aquaculture and other development areas. The LVFO, in conjunction with FAO, is working on a strategy and action plan to develop aquaculture in East Africa. The LVFO works hand-in-hand with institutions dealing with environment, water, land use, research, training, private sector, fish processing and export; Non-Governmental Organisations and Inter-Governmental Organisations working around the Lake. The LVFO work has promoted community participation in co-management through the Beach Management Units (BMU) in all the partner states around the Lake.

Another initiative of the LVFO has been the coordination with government institutions in partner states to develop and manage fisheries resources and implement them at village, sub-country, divisional, national and finally at regional level. These institutions are: the Ministry of Livestock and Fisheries Development, Kenya, the Ministry of Natural Resources and Tourism, Tanzania and the Ministry of Agriculture, Animal Industry and Fisheries, Uganda. With regard to management of fisheries, these ministries are charged with formulating fisheries legislation and ensuring implementation. Each partner state is to facilitate an increased and sustainable fish production and proper management of wild fish stocks.

The LVFO partners with the Kenya Marine and Fisheries Research Institute (KMFRI), Tanzania Fisheries and Research Institute (TAFIRI) and the National Fisheries Resources Research Institute (NAFIRRI) in Uganda to manage the resources of Lake Victoria. These research institutes work to establish acceptable fishing methods and gear, fish stock biomass, composition distribution and population structure especially of the commercially important fish stock. The institutes give information on the lake productivity process including production of food for fish and work to remove the Lake's pollutants and contaminants. They collaborate closely with fisheries management institutions to design and maintain fisheries data bases and develop fisheries socio-economics and marketing. Nevertheless, there are still issues to be researched on the Lake biota and fisheries, particularly the Nile perch. The Nile perch, *lates niloticus centropomiddae*, was introduced into Lake Kioga in 1955 and within some years, it was found in Lake Victoria. It is now the backbone of the lake's socio-economic activities with an estimated annual catch of 300,000 to 500,000 tonnes.[12]

Despite all these achievements, the LVFO faces the problem of balancing sustainable development with management of the fisheries resources. The situation is further complicated by the poverty of the fishing community. Lack of proper policies and legal framework guiding fishing activities has led to over-fishing of the Nile perch and other fish species which has led to cross-border conflict in some cases.

Other challenges include incidences of diseases among fishing communities such as HIV/AIDS, malaria and bilharzia. Poor agricultural practices have led to soil erosion and siltation in the lake whereas pollution from the industries, municipal and domestic waste contribute to the deterioration of water quality. The LVFO also has an inadequate statistics data collection, storage and dissemination system and limited funds for fisheries management and development and scientific management of information. Lastly, member states continue to strain in their overall support for the LVFO.

The East Africa Community (EAC)
The East African Community is an Inter-Governmental Organisation that was established in 1997 through the signing of the Treaty for Establishment of the East African Community by the three initial partners: Kenya, Uganda and Tanzania. Rwanda and Burundi have since joined the community. The objective of the EAC is to develop policies and programmes that are aimed at widening and deepening co-operation in political, economic, social and cultural fields, research and technology, defense, security, legal and judicial affairs, for the benefit of the partner states.[13]

In connection to Lake Victoria, the EAC has autonomous managment institutions such as the Lake Victoria Basin Commission (LVBC), CASSOA, the Lake Victoria Fisheries Organisation (LVFO), the Inter-University Council of East Africa (IUCEA) and the East Africa Development Bank (EADB), "since Lake Victoria is very important to East Africa, EAC became identified as being able to play a harmonising role of all activities in Lake Victoria."[14] The EAC focuses on the development of the Lake Victoria region through the administration of the Lake and its various institutions. However, the riparian states' cooperation through the EAC and other institutions of the Lake Victoria mainly exists symbolically meaning that the integration process needs improvement in the economic, social and political relationship of the EAC countries.[15]

Under the Council of Ministers, a body known as the Lake Victoria Basin Commission (LVBC) was established and it is the administrative system for Lake

Victoria under the Treaty for Establishing the EAC so as to foster cooperation in management of this shared resource. Regional arrangements bring together senior officials of partner states from the ministries responsible for water, fisheries, environment, forests, tourism and wildlife. The Sectoral committee manages terrestrial ecosystems, aquatic ecosystems and deals with policy, legal, institutional and pollution issues. The national arrangement involves the Departments of Environment and ministries of Water, Lands and Environment.

The East African countries are actively seeking win-win opportunities in the management of Lake Victoria. However, if they are to cooperate better, they will need to tackle pertinent issues of trade, economic and regional integration which will help to accelerate progress towards efficiency in managing lake issues and the achievement of the long-term goals of the EAC.

Notes

1. Mumma, Constansia. (2007. PhD.dissertation). *Managing Transnational Water Conflict in Africa: With reference to the Nile, Lake Victoria, Kagera and the Nile Basin*: Univertsity of Limpsig, Germany. p. 87.

2. Orach, Meza, Faustino & Okurut, Tom. *"Background to the preparation of the second phase of the Lake Victoria Environmental Management Project LVEMP-II"*: Available at www.ftp.Fao.org/agl/ Kagera docs/1o.programme-prjects/reg-LVEMP.pdf > (Accessed on April 08, 2009).

3. Atieno, Mumma Constansia. (1999 M.A Thesis). *"The Lake Victoria Water Hyacinth: Its Implication for International Environmental Conflict (IECS), Management and Regional Relations in East Africa"*. Institute of Diplomacy and International Studies (IDIS), University of Nairobi. p. 50.

4. Mumma, Constansia, p. 88.

5. Wambede, John. *"Progress in the Implementation of LVEMP-I and the Preparation of LVEMP-II"*: Available at www.Iwlearn.org/publications/misc/presentation/LVEMP%201%20 and%20 LVEMP%.22 (Accessed on April 08, 2009).

6. Wambede, John. *"Progress in the Implementation of LVEMP-I and the Preparation of LVEMP-II"*.

7. Lake Victoria Fisheries Organization, Strategic Vision for Lake Victoria (1999-2015). Jinja: LVFO Secretariat, 1999, p. 4.

8. Lake Victoria Fisheries Organization, Strategic Vision for Lake Victoria (1999-2015)

9. Lake Victoria Fisheries Organisation (LVFO): Available at www.lvfo.org (Accessed on April 08, 2009).

10. Lake Victoria Fisheries Organisation (LVFO). p. 5.

11. Atieno, Mumma Constansia. (1999 M.A Thesis). *"The Lake Victoria Water Hyacinth: Its Implication for International Environmental Conflict (IECS), Management and Regional Relations in East Africa"*. Institute of Diplomacy and International Studies (IDIS), University of Nairobi. p. 114.

12. Oyugi, Aseto et al. 2003. *Lake Victoria (Kenya) and it's Environs: Resources, Opportunities and Challenges*. Kisumu: Osienala. p. 21.

13. *East African Community: An Overview of Lake Victoria Basin Commission*. Kisumu, p. 21.

14. *East African Community: An Overview of Lake Victoria Basin Commission*. p. 154.

15. Swedish International Development Agency (SIDA), *Lake Victoria Region: Strategic Conflict Analysis*, Department for Africa, May 2004. p. 23.

National Waters

Kenya National Water Policy

Silas Mutia M'Nyiri

Introduction

Policy is a set of principles which is used as a basis for making decisions to further certain objectives. Almost any institution, whether public or private, operating for profit or voluntary, requires a policy to guide its operations and provide a frame of reference for its members. Ideally, a public policy is to be codified in the form of a written policy statement which has been formally endorsed by a body with the requisite authority (such as, in the case of a national policy statement, the cabinet). Particularly in the water sector, it is desirable that non-state actors be involved in the formulation of policy. This ensures that the policy is adapted to the circumstances prevailing in the country and that people will be more aware and more committed to ensuring that the intentions enunciated in the policy statements are in fact implemented.[1]

In many cases, policies are not codified in this *de jure* way. What actually happens in practice in the management of water can be analysed to deduce a *de facto* policy which may differ from what has been written or from what has been stated by government ministers or others who seek to enunciate water policies. Policy is also implicit in legislation. Ideally, the national water law will provide the legal framework for the implementation of national water policy, but again there may be observable differences between articulated policy, the codified legal framework and what is done in practice.[2]

These differences are not surprising or entirely undesirable. As circumstances, national aspirations and the dominant ideological framework change, so must the water policy change. A revision of policy may well start with an agreed water policy becoming less applicable and the level of adherence to the policy decreasing. To fill the gap, informal policy statements are made and debated. At some stage, a full-blown discussion of water policy becomes necessary, leading eventually to a new water policy statement, revised water law and new water sector institutions.

The initial water sector reform that took place after independence started following the launching of a policy document "Sessional Paper No. 10 of 1965 on African Socialism and its Application to Kenya". This paper directed the Government's policy towards priority areas for the African population, which were identified as poverty, illiteracy and diseases. The policy required that the core infrastructure for economic and social activity be in Government hands. Accordingly, the Government was engaged in all productive activities, including the provision of water and sanitation services, often at minimal charge to the consumer. In addition, the Government undertook programs to provide land to the people and some forest conservation areas were earmarked for human settlement. This situation undermined the sustainability of the water resource base.

In 1974, owing to the growing involvement of the Government in the development of water and sanitation services, the Water Department, under the then Ministry of Agriculture, was elevated to a full Ministry of Water Development. The Ministry intensified the Government's ambitious water development programme and envisioned achieving the provision of water for all by the year 2000. Consequently the Government became involved in complete management of almost 100 urban water supplies and 600 rural water supplies. After some time it was realised that the Government was not the best placed institution to undertake the role of water supply and sanitation provision. Water resources management was not regarded as a priority during this time.[3] The First National Master Water Plan (NMWP) was an excellent study which laid the foundation for the subsequent water development project implemented in the 1980 to 1990 decade.

Between 1990 and 1992 the Government undertook the Study on National Water Master Plan in collaboration with Japan International Cooperation Agency (JICA). The objective of the study on the National Water Master Plan, 1992 was to propose a nationwide framework of orderly planning and development of water resources in the country. This plan recommended for formulation of a water policy which culminated in the policy document in the water sector that was published in 1999 as Sessional Paper No. 1 of 1999, the "National Policy on Water Resources Management and Development."

In order to implement the policy, the then Ministry of Environment and Natural Resources begun by reviewing the Water Act (Cap 372) to spearhead the reform process. A new legislation, Water Act 2002 was enacted and came into operation on 18 March, 2003. It has provisions that would allow for the necessary reforms required for improved water resources management in Kenya to be achieved. Most importantly, the Water Act 2002 provides mechanisms for financing water resources protection and management. The Water Act 2002, therefore, enables the Ministry to implement the National Water Policy and establishes a new order of dispensation in the water sector amongst others.

This is timely given the fact that water crises experienced in the late 1990s generated serious public and private debate over water and its vulnerability to climate change and environmental degradation. For example, the *El Nino* of 1997/1998, the *La Nina* drought 2000/2001, and the severe droughts in 2007 and 2009, impacted all segments of society, thereby, highlighting serious weaknesses in existing water management systems. This weakness is further shown by the encroachment of water catchments for logging, settlement, farming or replacing indigenous trees with exotic trees such as Eucalyptus among others that transfer high volumes of water, thereby causing rivers to dry up.

The First two National Water Strategies namely the Water Resources Management Strategy and the Water Services Strategy give the road map for de-centralisation of operational activities from the central government to regional autonomous public bodies and other actors such as community, private sector, NGOs and National Water Conservation and Pipeline Corporation among others. It has also addressed the issues of the institutional framework and financing of the sector.

Additionally, the Environmental Management and Coordination Act of 1999, being the umbrella legislation that guides and coordinates activities with impacts on the environment under all other legislation touching on the management of natural resources, is now in place. All these documents provide an opportunity for implementing the reforms and from now on the water resource management shall be based and guided by the above legislations and policy direction.

Trans-boundary Water Management

The National Water Policy under Sessional Paper number 1 of 1999 in itself is silent on trans-boundary waters and therefore does not give any policy direction on shared water resources. It is therefore, the absence of policy direction in the National Water Policy that necessitated the Kenyan Government in 2002 to form a National Taskforce to formulate the Kenyan Policy on the utilisation of the Lake Victoria Water resources which is the major trans-boundary water resource in Kenya that harbours over 50% of the country's national water resources. The National Taskforce, after wide consultation with major stakeholders, came up with a Final Draft National Water Policy on the utilisation of the Lake Victoria water resources. Though this policy is widely accepted by the stakeholders, it is yet to be tabled in Parliament for adoption which is necessary as a mark of the final stage of policy formulation in Kenya. The key challenge now is to lobby the parliamentarians to accept and adopt this policy. The other challenge is to have this policy harmonised with the yet to be concluded Nile River Basin Cooperative Framework.

Though this draft National Water Policy on the Utilisation of the Water Resources of Lake Victoria has not been adopted by Parliament, it is used as a guide by the Kenyan Team to the Nile River Basin Cooperative Framework negotiations. This Draft Policy is also in harmony with the protocol for the sustainable development of the Lake Victoria Basin.

The Water Basins that Kenya shares with its neighbouring countries are as follows with:

Ethiopia: Kenya shares the water resources of the Omo River that flows in Lake Turkana, in addition to the Daua River which flows along the Kenya-Ethiopia border before entering Somalia.

Tanzania: Kenya shares Umba River, Lakes Jipe and Chala, Lumi River, Mara River and Lake Victoria.

Somalia: Kenya shares the Daua River and the Mert Aquifer.

Uganda: Kenya shares the Sio/Malaba and Malakisi Rivers in addition to Lake Victoria.

The Ministry of Water and Irrigation is almost finalising the trans-boundary water policy document for the management of these shared water resources. Bilateral Joint Management Riparian Commissions are also being proposed by Kenya to its neighbours, with whom they share these water resources.

Water Resources Management Challenges

Water Scarcity

Kenya is classified as a water-scarce country. The natural endowment of renewable freshwater is currently about 21 billion cubic metres (BCM) or 647 m^3 per capita per annum. A country is classified as "water scarce" if its renewable freshwater potential is less than 1,000 m^3 per capita per annum. By 2025, Kenya is projected to have a renewable freshwater supply of only 235 m^3 per capita per annum. This is likely to be even less if the current destruction of water catchments and the subsequent loss of water resources is not checked.

Only some 40% of the renewable freshwater has potential for development representing the safe yield. The remaining 60% are required to sustain river flows hence ensuring sustenance of ecological biodiversity. This 60% also acts as reserve for future development and maintains the trans-boundary waters' flow. Kenya's safe yield of surface water resources has been assessed to be 7.4 and 1.05 BCM per annum for surface and ground waters respectively. This kind of water situation poses a great challenge to water policy implimentation especially with regards to satisfying all the stakeholders.[4]

TABLE 1: RENEWABLE WATER RESOURCES BY DRAINAGE AREAS IN BCM/YR

Drainage Area	Catchment Area of the Drainage Basins km2	Total Annual Mean Surface Runoff (106 m3)	% of Annual Mean Surface Runoff by Drainage Basins	Groundwater (106m3) Groundwater potential by Drainage Basin	% of Groundwater potential by Drainage Basin	% of Total Water Resources potential by Drainage Basin
Lake Victoria	46,229	11,672	59.2	115.7	18.7	54.1
Rift Valley	130,452	2,784	1.0	125.7	20.3	3.4
Athi River	66,837	1,152	2.9	86.7	14.0	4.3
Tana River	126,026	3,744	33.5	147.3	23.8	32.3
N.Ewaso Nyiro	210,226	339	3.3	142.4	23.0	5.8
Total	579,770	19,691	100	610	100	100

Source: (National Water Master Plan, 1992 and National Water Master Plan After Care, 1998)

According to the study on National Water Master Plan 1992 by JICA and the aftercare of 1998, the average annual water available is 20.2 Billion Cubic Meters (BCM) distributed as shown in Table 1.

Lack of Co-operative Frameworks to Manage Shared Water Resources

Kenya shares about 54% of its water resources with the neighbouring countries and provides about 45% of surface water inflows to Lake Victoria and hence to the upper White Nile River. This inter-dependence between Kenya, its immediate neighbours, and downstream and upstream Nile Basin States has considerable implications for the management of the country's major water resources. Therefore, these resources must be jointly managed within agreed frameworks such as the Nile Basin Cooperative Framework, currently under negotiation, to avoid mistrust and tension and to allow for equitable and reasonable utilisation of the shared water resources for the benefits of all riparian States.[5]

Poor Prioritisation

Before the recent Water Sector Reforms, the Ministry of Water focused on water provision and sanitation services while little attention in terms of financial resources allocation was given to the water resources management and development. This wrong priority in the past has resulted in lack of adequate water storage capacity that today stands at 4 m^3 per person per year in addition to deteriorated water quality and diminishing quantities of water resources.

Due to the same poor prioritisation and lack of resources management organisation, the tax payer's money in Kenya is rarely spent on projects that could have a direct benefit for the ordinary Kenyans, like providing them piped water, or protecting the dwindling water resources for the common good.

Most Kenyans suffer from lack of water despite having big rivers flowing across their residential areas or having freshwater lakes next to them, without mentioning the flood waters that can be harvested and stored for use during dry seasons.

Inadequate Financial Resources

Other obstacles include inadequate financial resources to fully operationalise the new water institutions, and inadequate acceptance of the changes in the water sector as the changes are seen from some quarters as loss of existing opportunities. The newly - established institutions, though recent, have made significant gains in the water sector.

Lack of Public-Private Partnership

Though there is generally political good will for the water sector reforms in Kenya, lack of clear policies on a public-private partnership with regard to private sector participation in the provision of water supply and sanitation services presents another challenge.

Degradation of Water Catchments

Weak policy formulation and enforcement, political factors, macro-economic policies (Cash crop production for export) and population pressure among other factors have resulted in inappropriate land use changes leading to catchments' degradation, poverty, soil erosion, degradation of both quantity and quality of water resources, deterioration of riparian lands causing flash floods, turbidity and siltation of watercourses and storage facilities. This is best illustrated by the Mau Forest degradation case study discussed below.[6]

Mau Forest Degradation Case Study

Mau forest blocks include: Ol Pusimoru, Eastern Mau, Maasai Mau, Western Mau, South West Mau, Transmara and Eburu Catchments. The Mau Complex has the largest block of closed - canopy forest in East Africa and provides direct and indirect benefits in terms of livelihood to nearly 15 million people both in Kenya and Tanzania. It forms the largest water tower - Mau forests form the upper catchment of Kenya's main rivers – the Sondu, Yala, Nzoia and Nyando rivers all flowing into Lake Victoria, as well as the Ewaso Ngiro, Kerio and Mara rivers. It also supplies water to many lakes in the Rift Valley, from Lake Turkana bordering Ethiopia to Lake Natron in Tanzania.

The regular breeding site for the more than two million flamingos is found on the Rift Valley lakes of Eastern Africa with Lake Nakuru hosting the largest population of flamingo world-wide. Further, Mau complex supports a great deal of the national economy including power generation, tea sector (main tea estates in Kericho), tourism and wildlife. The role of this complex in supporting the Masai Mara Game Reserve and its key function in the survival of wildlife in Masai Mara Game Reserve and Serengeti National park in Tanzania cannot be overemphasised. Sondu Miriu Hydro Power Complex which will contribute about 60 MW to the national grid derives its water from Mau complex. Despite these water related benefits of the Mau forest, extensive deforestation of the complex has taken place due to the following factors:

a) Plantations

In 1930, the white settlers cleared parts of Mau forests for the establishment of forest plantations using mainly exotic species. These occupy about 10% of the forest. The Ogiek community have been receding into the natural forests when this happens as their lifestyle depends on natural forests. In recent times, most of these plantations found surrounding the indigenous forests have been cleared to pave way for agriculture. Without the protection offered by the plantation, the indigenous forests are now threatened.

b) Logging

Saw millers obtain licenses to permit them to practice logging especially within the plantations. They also have to pay logging fees to the Forest Department. These charges are very low and not revised often to reflect the current economic situation. The forests have been logged extensively in a non-systematic way. Often, there is intensive selective cutting of trees and overexploitation leaving behind an inferior stock to mature as a final crop. This leads to further forest loss as the Forest Department cannot rehabilitate the areas where trees have been felled.

c) Human Settlement

The forest is increasingly being cleared for human settlement as the government is keen to settle the forest - dwelling communities along the forest boundaries or within the cleared forest plantations.

d) Fires

In order to clear land for cultivation or grazing, those intending to settle into the forest set the vegetation on fire. The fires spread extensively and cause a lot of damage to the forest biodiversity. Charcoal burners also destroy the forest as they use traditional kilns which are not energy efficient. This activity is done illegally and therefore no one controls or tends the fires which in most cases destroy the areas surrounding the charcoal burning sites.

e) Forest Excisions

The process of forest protection was introduced under the 'East Africa Forest Regulations, 1902' by the first Conservator of Forests. These regulations allowed for the gazettement and degazettement of forests, and control of forest exploitation through a system of licenses and fines. The Government, through the Minister for Natural Resources has the express authority to degazette the forest through a legal process of excision. These excisions are done with the intent of converting the area to other alternative land uses like settlement, and private agriculture, which do not foster tree cover. The forests are degazetted then surveyed and demarcated for the proposed use.

The forest excision process has several loopholes that include the excisions being made without consultation with the stakeholders. Procedures of collecting public views and sharpening their perspectives on causal-effect linkages as it may affect those aggrieved by the excision are never put in place and neither are provisions for compensation clear. A notice is placed in the *Kenya Gazette* and whoever wants to contest it is given 28 days to do so. The Minister is however under no obligation to consider the views in the final decision. The readership of the Kenya Gazette is very limited and not many people get to know about the notices.

The Minister has the powers to put in a notification and he can be influenced by political and economic pressure but not necessarily for the common interest of the public. Excisions usually take place after the forests have already been illegally occupied. There is no environmental and socio-economic impact assessment done for the proposed changes in land-use leading to unsustainable land management.

Impact of Mau Destruction

The destruction of Mau Forest has resulted in:
• Drastic reduction of the former perennial rivers into seasonal rivers, e.g. the ones that flow into Lake Nakuru;
• Recession of Lake Nakuru which is a Ramsar site and hosts international birds. As a result of this, movement of birds from the lake is reported;
• Increased pollution of Lake Nakuru due to decreased dilution factor as a result of low flows into the lake has led to the death of flamingos;
• Increased workload and longer distances to water sources for women and girls posing a time, health, personal and food security threat;
• Increased migration or longer distance to water points for men in pastoralist/ dry areas. This is a double tragedy for women, girls and boys whose education is interrupted during migration;
• Loss of productive opportunities especially for women – school opportunities for girls are lost as they have to engage in water - fetching activities for their family water needs;
• Poor sanitation leading to increased school drop-out rates for girls;
• Water, sanitation and hygiene related sicknesses are very common where water is scarce;
• Health issues have implications on women's time because they have to cater to the sick; and
• Increased poverty as family budgets tend to focus on treating water-related sicknesses.

The National Water Policy Objectives

As already discussed above, a public policy is codified in the form of a written policy statement which has been formally endorsed by a body with the requisite authority. Particularly in the water sector, it is desirable that non-state actors be involved in the formulation of policy. This ensures that the policy is adapted to the circumstances prevailing in the country and that people will be more aware and more committed to ensuring that the intentions enunciated in the policy statements are in fact implemented.

A revision of policy may well start with an agreed water policy becoming less applicable and the level of adherence to the policy decreasing. To fill the gap, informal policy statements are made and debated. At some stage, a full-blown discussion of water policy becomes necessary, leading eventually to a new water policy statement, revised water law and new water sector institutions.

In the current context, the *purpose* of a water policy statement or document is to establish principles of equitable, efficient and sustainable utilisation of water resources. Moving from document to implementation, the *purpose of the water policy* itself is to maximise the economic and social benefits of water while ensuring these are shared in an equitable manner and that environmental sustainability is preserved.[7]

Pursuant to this, our national water policy was drawn with the following objectives:

- Preserve, conserve and protect available water resources;
- Supply water of good quality and in sufficient quantities for various needs;
- Establish an efficient and effective institutional framework for the water sector;
- Develop a sound and sustainable financing system for effective and efficient water resources management, water supply and water - borne sewage collection, treatment and disposal;
- Treat water as a social and economic good;
- Sustainable, rational and economical allocation of water resources;
- Ensuring safe wastewater disposal for environmental protection.[8]

This Water Policy provides strategies that aim at achieving sustainable development and management of the water sector by providing a framework in which the desired targets/goals are set, outlining the necessary measures to guide the entire range of actions and to synchronise all water - related activities and actors. The basic areas the policy has addressed itself to include water resources management, water supply and sewerage development, institutional arrangement and financing of the water sector. It is intended to bring about a culture that promotes comprehensive water resources management and development with the community participation as the prime mover in the process to guarantee sustainability. This strategy would ensure that the Government's role would be largely to provide policy guidelines for the sector.[9]

Institutional and Legal Framework

The present institutional arrangements for the management of the water sector in Kenya can be traced to the launch in 1974 of the National Water Master Plan whose primary aim was to ensure availability of potable water, at reasonable distance, to all households by the year 2000. The Plan aimed to achieve this objective by actively developing water supply systems. This necessitated that the Government directly provide water services to consumers, in addition to its other roles of making policy, regulating the use of water resources and financing activities in the water sector. The legal framework for carrying out these functions was found in the then prevailing law, the Water Act, Chapter 372 of the Laws of Kenya.

In line with the Master Plan, the Government upgraded the Department of Water Development (DWD) of the Ministry of Agriculture into a full Ministry of Water. DWD embarked on an ambitious water supply development programme. By the year 2000, it had developed, and was managing, 73 piped urban water systems serving about 1.4 million people and 555 piped rural water supply systems serving 4.7 million people.

In the 1980s the Government begun experiencing budgetary constraints and it became clear that, on its own, it could not deliver water to all Kenyans by the year 2000. Attention therefore turned to finding ways of involving others in the provision of water services in place of the Government, a process that came to be known popularly as "handing over." The Government developed a fully - fledged policy, The National Water Policy, which was adopted by Parliament as Sessional Paper No. 1 of 1999. The Policy also stated that the Water Act, Chapter 372 would be reviewed and updated, attention being paid to the transfer of water facilities.

While developing the National Water Policy, the Government also established a National Task Force to review the Water Act, Chapter 372 and draft a Bill to replace the Water Act, Chapter 372. The Water Bill 2002 was published on 15 March, 2002 and passed by Parliament on 18 July, 2002. It was gazetted in October 2002 as the Water Act, 2002 and went into effect in 2003 when effective implementation of its provisions commenced.

Water Sector Reforms

The Water Act 2002 has introduced comprehensive and radical changes to the legal framework for the management of the water sector in Kenya. These reforms revolve around four pillars that include:

• The separation of the management of water resources from the provision of water services;

• The separation of policy making from day to day administration and regulation;

• Decentralisation of functions to lower level state organs; and

• The involvement of non-government entities and communities in form of Water Users Associations in the management of water resources and in the provision of water supply and sanitation services.10

The implementation of the water sector reforms has so far been very impressive with a few challenges such as sustainability of the new institutions due to financial inadequacies. The institutional framework resulting from these reforms is represented diagrammatically in the Figure below.

Diagrammatoic representation of the new institutional structure for the management of water affairs in Kenya

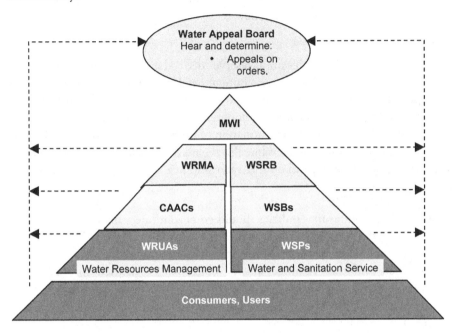

Source: Ministry of Water and Irrigation, Transfer Plan, 2004.

Catchment Management Approach

Sustainable water resources management requires the adoption of an ecological approach. This can be achieved by the establishment of a minimum flow which must be maintained at all times. Sustainability requires also that policies should go further and strive to maintain selected rivers with high ecosystem functions and values in their natural state. The Water Act 2002 does this through three mechanisms for managing the use of water resources: basin wide or catchment management, the concept of "the reserve" and "water quality objectives".

Kenya's laws on the management of water resources have traditionally been designed on the basis of catchments. Under the Water Act, Chapter 372 (now repealed) the country was divided into six catchments, as the basis for water resources management. In practice, catchment management has been made difficult by the Government's policy commitment to, and implementation of, an administrative system based on districts, as outlined in the policy paper on District Focus for Rural Development Strategy of 1986.

The Water Act, 2002 has maintained the catchments-based approach to the management of water resources. It establishes the Water Resources Management Authority (WRMA) which is required to designate catchment areas, and to formulate for each area a catchment management strategy. Though still at its formative stages, the Government direction this time round is actually to decentralise management of water resources to the catchments, as required by the law, and moving away from current decentralisation of management to the districts under the district focus policy.[11]

The Reserve

The Water Act 2002 provides for a "reserve" defined as "the quantity and quality of water required to: (a) satisfy basic human needs for all people who are or may be supplied from the water resource, and (b) protect aquatic ecosystems in order to secure ecologically sustainable development and use of the water resource."[12] It must be emphasised that the determination of the reserve must not only take into account ecological considerations but also the need to maintain the livelihoods of the communities that depend on the water resource.

Water Quality Objectives

Sustainability requires that policies should go beyond mere maintenance of "the reserve". An effort should be made to preserve selected rivers with high ecosystem functions and values in their natural state. This requires that water resources be classified and that resource quality objectives be defined for each class of water resource. Doing so opens the way for a defined category of water resources to be classified as of high ecosystem value deserving to be maintained in their natural state.

Ministry of Water and Irrigation (MWI)

The Ministry of Water and Irrigation (MWI) was established in January 2003 through the Presidential Circular No. 1/2003. The creation of the Ministry consolidated the responsibility for the management and development of water resources in one docket. The mandate of the Ministry is "to protect, harness and develop potential of water resources to ensure availability of quality water to all". Given this mandate, the vision of the ministry has been stated as *"Assured water resources availability and accessibility by all"* while the mission is *"To promote integrated water resources management and development through stakeholder participation to ensure water availability and accessibility to enhance national development".[13]*

The core functions of the ministry have been articulated as.[14]:

- The conservation, control and protection of water catchments areas;
- The review, documentation and implementation of water resources and water apportionment policies;
- Assurance of water quality and pollution control;
- Rural and urban water development and supply;
- Flood control and land reclamation;
- Through the National Irrigation Board, the ministry has the mandate to undertake irrigation projects across the country;
- Wastewater treatment and disposal.

The new water institutions established under the Water Act 2002 have been set up and they include:

Water Resources Management Authority (WRMA)

The role of the WRMA includes implementation of policies and strategies relating to management of water resources; development of catchments level management strategies including appointment and facilitation of Catchment Areas Advisory Committees (CAACs). Within the regional framework of the WRMA, catchment areas committees will advise on water resources conservation, use and apportionment; grant adjustment, cancellation or variation of water permits, etc. The establishment and operation of Water Resources Users Associations shall be encouraged and facilitated as fora for conflict resolution and co-operative management of water resources in catchment areas. The Water Resources Management Authority was established on 18 March, 2003 under section 46 of the Water Act 2002 as a State Corporation. The Authority offices are currently housed under the Ministry of Water and Irrigation, Nairobi.[15]

Water Services Regulatory Board (WSRB)

The role of this board includes overseeing the implementation of policies and strategies relating to provision of Water and Sanitation Services; regulating the provision of Water and Sanitation Services; licensing Water Services Boards (WSBs) and approving their appointed Water Services Providers (WSPs) through Service Provision Agreements; setting rules, establishing standards guidelines, and monitoring the performance of WSBs, WSPs and enforcing regulations. Establishing technical, water quality and effluent disposal standards. The Water Services Regulatory Board was established on 18 March, 2003 under section 46 of the Water Act 2002 as a State Corporation.[16]

Water Services Board (WSB)

As a Licensee, the WSB shall be responsible for the efficient and economical provision of water and sanitation services within its area of jurisdiction. Seven WSBs have been gazetted and established to cover the entire country. In addition, the Services Boards shall be the water asset holders and managers.[17]

Water Services Providers (WSP)

Direct provision of water services shall be undertaken by Water Services Providers (WSPs) who shall be agents of WSBs except where the WSRB is satisfied that the procurement of such agents is not possible or that provision of services by such agents is not practicable. These may be community groups, Non-Governmental Organisations (NGOs), autonomous entities established by Local Authorities or the private sector.[18]

Water Services Trust Fund (WSTF)

The WSTF assists in the financing of provision of water supplies in areas that are inadequately covered.[19]

Water Appeal Board (WAB)

The WAB adjudicates disputes within the water sector.[20]

Kenya Water Institute (KEWI)

The Kenya Water Training Institute (KEWI) is a middle training institute mandated to train water technocrats. The institute collaborates with National Public Universities in carrying out applied water research. KEWI therefore, could be considered as an institution for short - duration training in trans-boundary water resources management.[21]

Notes

1. Alan, Nicol. 2006. *Water Policy Guidelines and Compendium of Good Practice*. Nairobi.

2. Alan, Nicol. *Water Policy Guidelines and Compendium of Good Practice*.

3. Rafik, Hirj, Francois Marie & Deborah, Rubin. Seminar Proceedings on "Integrated Water Resources Management in Kenya", Nanyuki, 1996.

4. Seminar Proceedings on "Integrated Water Resources Management in Kenya", Nanyuki, 1996.

5. Nyaoro, John. 2005. *Final Situation Analyses on the National Water Policy in Kenya*, Nairobi.

6. Nyaoro, John. 2005. *Final Situation Analyses on the National Water Policy in Kenya*.

7. Nyaoro, John. 2005. *Final Situation Analyses on the National Water Policy in Kenya*.

8. Ministry of Water Development: National Water Policy 1999, Nairobi.

9. Nyaoro, John, *Final Situation Analyses on the National Water Policy in Kenya*. Nairobi, 2005.

10. Water Act 2002, Nairobi.

11. Ministry of Water and Irrigation: *The National Water Resources Management Strategy* (NWRMS), 2005 – 2007. Nairobi.

12. Water Act 2002, Nairobi.

13. Ministry of Water Development: National Water Policy 1999, Nairobi.

14. Ministry of Water and Irrigation: *Plan For the Transfer of Management and Operation of Water Services to Water Services Boards*. Nairobi, 2004.

15. Ministry of Water and Irrigation: *National Water Resources Management Strategy, 2005 – 2007*.

16. Ministry of Water and Irrigation Transfer Plan, 2004.

17. Water Act 2002, Nairobi.

18. Water Act 2002.

19. Water Act 2002.

20. Water Act 2002.

21. Nyaoro, John. 2005. *Final Situation Analyses on the National Water Policy in Kenya*.

Water Sector Reforms in Kenya: Institutional Set-up, Impact and Challenges in Urban Water Supply

Samuel O. Owuor

Introduction

Like other countries in sub-Saharan Africa, Kenya's socio-economic development goals are highly dependent on the availability of water in good quantity and quality. The government's long-term objective is to ensure that all Kenyans have access to clean potable water, and that water is available for key economic activities. In addition, it recognises that for the country to meet its poverty-eradication strategies and achieve the Millennium Development Goals (MDGs) water has to be made available, accessible and affordable, especially to the poor. This is based on the fact that all the eight MDGs are directly or indirectly related to access to water. The water sector reforms now being implemented in Kenya under the Water Act 2002 of the Laws of Kenya are designed to contribute to the realisation of this long-term objective as well as to addressing the policy, regulation and service provision weaknesses in the previous Water Act Cap 372. This chapter will present the institutional set-up of the water sector reforms in Kenya and thereafter, based on an inventory survey of five towns (Eldoret, Kisumu, Homa Bay, Kisii and Nakuru), discuss the emerging impact and challenges of the sector reforms in urban water supply.

Background to Water Sector Reforms in Kenya

Water governance has been identified as a key issue in water resources management as well as water services delivery, especially in sub-Saharan Africa.[1] The first attempt to 'reform' the water sector came as early as 1974 when the first *National Water Master Plan* was launched.[2] The primary aim of the Plan was to ensure availability of potable water, at a reasonable distance, to all households by the year 2000 – under the legal framework of Water Act Cap 372. In line with the Plan, the government upgraded the Department of Water Development of the Ministry of Agriculture into a fully - fledged Ministry of Water to coordinate actors involved in the provision of water and sanitation services.[3] However, the Ministry lacked financial resources and the Plan was not sustained. As the needs of the country changed over time, there were various government policy pronouncements. Among them was the *Sessional Paper No. 1 of 1986 on Economic Management for Renewed Growth* from which the government spelt out strategies for provision of basic services and reforms necessary to accelerate economic growth.[4]

In 1998 the government established the National Water Conservation and Pipeline Corporation (NWCPC) to take over the management of government - operated water supply systems that could be run on a commercial basis. In addition, large municipalities were allowed to supply water within their areas. Also allowed to operate were a number of donor-funded or supported community self-help water supply projects.[5] Although nominally autonomous with the opportunity for commercial orientation, NWCPC failed to attain financial viability or to improve the provision of water supply as originally envisaged. Neither could the local authorities do any better.

The idea of water sector reforms in Kenya gained momentum (again) in 1999 following the publishing of the *Sessional Paper No 1 of 1999 on National Policy on Water Resources Management and Development*. The paper identified and analysed the shortcomings in water resources management, water and sewerage development, institutional framework and financing of the water sector.[6] In other words, the weaknesses in policy, regulation and service provision in the previous set-up are the main drivers towards water sector reforms. These weaknesses, which the sector reforms intend to address, are summarised in Table 1.[7]

With the adoption of the Water Act 2002, all Kenyan municipalities are obliged to reform their water services along 'business' lines. The key word is 'commercialisation': water is not only a social good but also an economic good and water services have to be managed "in accordance with sound business principles".[8] Sections 11(1) and 11(2) of the Act laid the foundation for the *National Water Resources Management Strategy*.[9] The overall goal of the Strategy is "to eradicate poverty through the provision of potable water for human consumption and water for productive use". In short, water is considered by the Kenyan government as both a social and an economic good, to be available for all Kenyans and at a price reflecting its market value (cost recovery). The government also recognises that the poor cannot afford to pay such prices, a problem that has to be solved by subsidised rates. Further, the government stresses the importance of involvement of all stakeholders – including consumers, and women in particular – in the management of the country's water resources.

TABLE 1: BOTTLENECKS IN THE WATER SECTOR UNDER WATER ACT CAP 372

Policy formulation	• Poor co-ordination in the sector • Poor policy accountability • Poor attention to water resources management
Regulation	• Lack of a clear regulatory framework • Lack of monitoring and evaluation • Poor performance of water-undertakers
Service provision	• Poor management of water resources (quality and quantity) • Failure to attract and retain skilled manpower • Inadequate allocation of resources • Poor, inefficient and unreliable service delivery • Low coverage of water supply and sewerage services • Inability to attract investments • Dilapidated water supply and sewerage infrastructure • High levels of unaccounted-for-water • Low revenue collection, including corruption

Source: Kenya (2006b)

The Institutional Set-up of Water Sector Reforms in Kenya

The Water Act 2002, which became effective on 18 March 2003, provides the legal framework for the implementation of the water sector reforms based on the following guiding principles:

• The separation of water resources management from water supply and sewerage services;

• The institutional separation of policy formulation, regulation and service provision functions.

Decentralisation, participation, autonomy, accountability, efficiency, affordability and sustainability. For example, (-) *decentralisation* of services to the regional and local levels, i.e. to the Water Services Boards, Water Service Providers, Catchment Areas Advisory Committees, and Water Resources Users Associations; (-) *participation* of all the stakeholders; (-) financial and operational *autonomy* of the Water Service Providers; and (-) financial and ecological *sustainability* in the management of water resources. Institutionalising support to the financing of water services for under - served areas, i.e. the Water Services Trust Fund. Establishing mechanisms for handling disputes in the water sector, i.e. the Water Appeal Board.

Figure 1 presents the 'famous triangle' summarising the institutional set-up of water sector reforms under the Water Act 2002 while Table 2 is a summary of the roles and responsibilities of institutions in the sector reforms. The Act provides for the establishment of three levels of institutions for the provision of water supply and sewerage services: Water Services Regulatory Board (WSREB), Water Services Boards (WSBs), and Water Service Providers (WSPs). On the other hand, the management of water resources is under Water Resources Management Authority (WRMA) and Water Resources User Associations (WRUAs). The

Water Resources Management Authority executes its mandate through the Catchment Areas Advisory Committees (CAACs) whose membership consists of government officials, stakeholders and the community. Two of these institutions – Water Services Boards and Water Service Providers – are further discussed below as they are directly concerned with the provision of water supply in towns.

Figure 1: The Institutional set-up of water sector reforms under Water Act 2002.

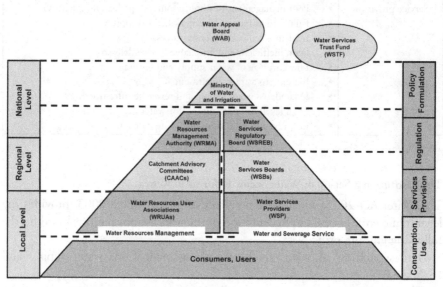

Source: MWI (2005)[10]; Kenya (2006b)[11]

It is expected that the clear roles and responsibilities defined to sector actors will result in improved water sector performance. At the *policy formulation level*, the sector reforms are expected to improve coordination in the water sector, enhance clear policy accountability, and give more attention to water resources management. At the *regulation level*, the sector reforms are expected to set in place a clear regulatory framework, enhance monitoring and evaluation, and improve performance of water undertakers. Lastly, the expected outcomes at the *service provision level* include improved management of water resources (quantity and quality), ability to attract and retain skilled manpower, improved and efficient service delivery, increased coverage, ability to attract investments, and improved infrastructure. Mumma[12] provides a clear legal interpretation and critical analysis of the Water Act 2002.

Water Services Boards (WSBs)

There are seven WSBs in Kenya: Athi Water Services Board, Tana Water Services Board, Northern Water Services Board, Coast Water Services Board, Rift Valley Water Services Board, Lake Victoria North Water Services Board and Lake Victoria South Water Services Board. WSBs were created to take full responsibility for the provision of water services through signing of Service

Provision Agreements with Water Service Providers. According to the Act, they are the legal owners of water and sewerage assets in their areas of jurisdiction. As such, they are responsible for the planning, development and expansion of water and sewerage services. They contract water and sewerage services provisions to Water Service Providers and monitor service delivery as well as having powers to lease assets, from their owners, for water service provision.[14]

Water Service Providers (WSPs)

The actual water service delivery to the consumers is done by the WSPs. The Act requires that a Water Services Board enters into a contract with a WSP through signing the Service Provision Agreement. In other words, the direct service providers are WSPs and not Water Services Boards. There are three categories of WSPs: (1) urban water service providers, which are incorporated as limited liability companies owned by one or more local authorities; (2) community water supplies which are managed by WSPs but registered as Water Resources User Associations by the Registrar of Societies and (3) private WSPs which include NGOs and private organizations.[15]

TABLE 2: ROLES AND RESPONSIBILITIES OF INSTITUTIONS IN THE SECTOR REFORMS

Institution	Roles and responsibilities
Ministry of Water and Irrigation	Policy formulation, sector coordination, monitoring, financing and supervision
Water Resources Management Authority	Regulation of water resources management through (-) developing principles, guidelines and procedures for the allocation of water resources; (-) assessing water resources potential; (-) determining and monitoring permits for water use; (-) regulating and protecting water resources; (-) determining water user charges and fees from source; and lastly (-) maintaining a database on water resources.
Water Services Regulatory Board	Regulation of water and sewerage services through (-) issuance and monitoring of licenses for the provision of water; (-) determining and monitoring standards for the provision of water services to consumers; (-) providing procedures for handling and dealing with complaints from consumers; and (-) developing tariff guidelines for the provision of water services.
Catchment Areas Advisory Committees	Advise the Water Resources Management Authority on issues concerning management of water resources at the catchment level.
Water Services Boards	Responsible for the efficient and economical provision of water services within their area of jurisdiction through signing of service provision agreements with Water Service Providers.
Water Resources Users Associations	Provides a forum for conflict resolution and cooperative management of water resources in designated catchment areas. In other words, it enables the public and communities to participate in managing water resources within their catchment areas.
Water Service Providers	Direct provision of water and sewerage services as agents of Water Services Boards.
Water Appeal Board	Handle disputes in the water sector.
Water Services Trust	Support financing of pro-poor water services in unserved areas.
National Water Conservation and Pipeline Corporation	Bulk water supply, dam construction, flood control, land drainage, ground water development and Ministry of Water and Irrigation reserve Water Service Provider.
Kenya Water Institute	Training and research

Source: MWI (2005)[13]

Under the Act, autonomous Water and Sanitation (or sewerage) Companies – WASCOs – are given the responsibility to provide water and sanitation services within urban areas. The lead partners in this venture are the local authorities.

The WASCOs operate within the jurisdiction and oversight of the Water Services Boards, instrumental in their registration and incorporation. The WASCOs are expected to be managed on commercial principles, including signing performance contracts, cost-recovery, and sustainability within a context of efficiency, operational and financial autonomy, accountability and strategic, but minor, investments. They are supposed to improve access to water and sanitation services for poverty reduction and sustainable development. In fact the core mandate of the WASCOs is to provide effective, efficient, adequate and safe water to customers and to collect, treat and dispose sewage in a safe and environmentally friendly manner.

The Impact of Water Sector Reforms in Urban Water Supply

There is no doubt that the water sector reform has reported tremendous improvements. As indicated before, water sector reforms in Kenya are intended to address the weaknesses in policy, regulation and service provision characteristic in the previous Water Act Cap 327. The expected outcomes of the water sector reforms are summarized in Table 3. However, it is not the intention of this section to analyse in how far these expected outcomes have been achieved.

TABLE 3: EXPECTED OUTCOMES OF WATER SECTOR REFORMS

Policy formulation	Improved co-ordination in the water sector Clear policy accountability Focused attention to water resources management
Regulation	Clear regulatory framework Performance in monitoring and evaluation Improved performance of water-undertakers
Service provision	Improved management of water resources (quality and quantity) Ability to attract and retain skilled manpower Efficient provision of services leading to self sustainability Increased coverage Ability to attract investments Improved infrastructure

Source: Kenya (2006b)[16]

The following section presents the (emerging) impact of water sector reforms in urban water supply. The discussion is based on an inventory survey of water companies in five towns: Eldoret Water and Sanitation Company (Eldoret); Kisumu Water and Sewerage Company (Kisumu); South Nyanza Water and Sanitation Company (Homa Bay); Gusii Water and Sanitation Company (Kisii); and Nakuru Water and Sanitation Services Company (Nakuru).

Minimal Network Extension with Efforts Towards Rehabilitation and Water Kiosks

All the five water companies operate within their municipality boundaries, which are of different sizes. The intensity of coverage in the municipality is still based on the existing water supply network inherited from the local authorities, the National Water Conservation and Pipeline Corporation and/or the Ministry of Water and Irrigation. As such, it is common to find that the central business districts and the high-income neighbourhoods are better connected than, for example, other parts of the city. Generally, none of the towns is yet to achieve maximum coverage. While Eldoret and Nakuru boasts of 60% of the municipality being covered, less than half of Kisumu (40%) and Homa Bay (30%) municipality have access to piped water. However, the low-income estates are more often than not poorly connected or not connected at all. In short, service coverage is generally below target and expansion of the existing infrastructure is still minimal.

Whereas there is insignificant network extension, i.e. in terms of new pipelines, efforts are being targeted to rehabilitating the existing network by replacing the old pipes. Although it is not clear to what extent, Eldoret Water and Sanitation Company, Kisumu Water and Sewerage Company and Nakuru Water and Sanitation Services Company, indicated that they have increased their network coverage by adding new pipelines. In Langas, one of the largest informal settlements in Eldoret, the need for household connections has increased to the extent that some of the existing water kiosks have been rendered functionally redundant. This is because the situation of water supply in the low-income estates is largely addressed through the provision of water kiosks. This is what is happening in Homa Bay and Kisii where water kiosks are being constructed to supply water to the poor neighbourhoods as a short-term intervention.

Significant Reduction in Unaccounted-for-water

Unaccounted-for-water is the difference between the quantity of water supplied to the network and the metered quantity of water used (and paid for) by the customers. To a large extent, the level of unaccounted-for-water is an indicator of how well a utility is managed. A reduction in unaccounted-for-water means improved revenue and saving the scarce water resources. Reduction of unaccounted-for-water within the distribution system, efficient irrigation methods, recycling and re-use of water, and rainwater harvesting, including roof catchment for domestic purposes are some of the water conservation and demand management strategies that can be used in urban areas. It is no doubt that the water companies inherited high unaccounted-for-water at their inception, all above 70%.

In an effort to meet their performance targets, the water companies in the five towns have reduced their unaccounted-for-water to 62% in Kisumu, 54% in Kisii, 52% in Homa Bay, 43% in Nakuru, and 35% in Eldoret. Except for Eldoret and Nakuru, the other towns are still far from the recommended proportion of 25%. However, given the enormous challenges the water companies are facing as they implement the sector reforms, this reduction is indeed significant. Nakuru and Eldoret are doing comparatively better because they started water sector reforms much earlier. Specifically, they were among the three towns selected (together with Nyeri) to pioneer commercialisation of water supply through water and sanitation companies in Kenya.

The high unaccounted-for-water and its continued persistence are attributed to physical (or technical) losses and administrative (or commercial) losses. Physical losses occur largely through leakage brought about by the ageing pipes and storage tanks. Administrative losses result from illegal connections; lack of leak detectors; defective and non-functional meters; flat rate tariff due to lack of meters; inefficient, incorrect and false meter readings and billings; and wastage of water at communal water points as is the case in Nakuru's municipal council housing estates. The water companies have been able to address the high unaccounted-for-water in various ways. For example, creation of a department in charge of supervision; engaging private investigators and giving incentives to whistle blowers; quick response and repair to reported cases of leakages (in Kisumu); door-to-door impromptu checks for illegal connection (in Homa Bay); block mapping and awareness campaigns (in Kisii); and making use of new meters and leak detectors (in Nakuru).

Towards Improved Metering, Billing and Revenue

Although all the water companies alluded to the fact that metering, billing and revenue had improved, it is not possible at this stage to analyse by what proportion. However, there are indications that the companies are in the process of improving their metering, billing and revenue – albeit gradually. In Eldoret, 70% of the total connections pay their bills while in Kisii, half of the registered connections in the municipality are active. Kisumu's metering ratio is 100% with a high revenue collection efficiency of 90% (increasing from 50%). In Nakuru, metering has improved to 88% and in 2008 alone 12,000 meters were installed and connected to consumers. To improve metering, billing and revenue, the Lake Victoria Region Water and Sanitation (LVWATSAN) Initiative in Homa Bay and Kisii has provided the water companies with new meters.

Addressing the Plight of Low-income Neighbourhoods through Pro-poor Programmes

Despite the lack of a clear policy on pro-poor programmes, the water companies in the five towns recognise the need and importance of supplying water and sanitation services to the low-income neighbourhoods – where the large majority of the residents in these towns live. While Eldoret Water and Sanitation Company may be ahead in their pro-poor focus, other towns have also initiated a number of pro-poor programmes. Notable among them is the continued provision of water kiosks or standpipes in the low-income settlements. These water kiosks are supposed to serve a number of people in the neighbourhood as well as supplying safe and affordable water.

The 'delegated management model' being pioneered in Nyalenda – a densely populated slum area in Kisumu – is another example. The project is intended to increase access to safe and affordable water to the urban poor. A similar project in Kisumu's peri-urban low-income area but owned, operated and managed by the community is the Wandiege Community Water Supply Project. Eldoret has long-term plans to extend piped water supply to some of its low-income neighbourhoods while at the same time continuing to supply water through water kiosks. Despite the water kiosks, residents are still relying on other highly - priced and poor - quality sources of water, i.e. water vendors, wells, springs, etc.

Attraction of Donor Funding and Interventions

Water Services Boards are now able to attract and secure funding for rehabilitation and expansion of water and sanitation services. For example, the French government is active in Kisumu, UN-HABITAT in Homa Bay and Kisii, and African Development Bank in Nakuru. Some of these interventions, for example the LVWATSAN programme, have also brought new concepts in water governance. The multi-stakeholder forums in Homa Bay and Kisii are good examples of how various stakeholders, including the urban poor and women, are involved in water governance. However, it is not clear whether "water reforms seem to have been implemented hurriedly to impress the donors" as one of the respondents said or whether it was meant to attract donor funding and interventions.

Bobaracho Self-help Group Water Kiosk

Photo: Dick Foeken.

Providing an Opportunity for other Water Service Providers

The Water Act 2002 allows for other Water Service Providers as long as they have been registered and given permit to operate by the Water Services Board. An example in these towns is Wandiege Community Water Supply Project in Kisumu. The Wandiege water supply project, running along the same principles as Kisumu Water and Sewerage Company but on a much smaller scale, is wholly owned, operated and managed by the community. It supplies safe and affordable water to the people living around its water supply – a borehole.

The Challenges of Water Sector Reforms in Urban Water Supply

Public or Private Companies?

It is not clear whether the water companies are private or public limited companies. Whereas under the Companies Act, water and sanitation companies are registered as private, limited liability companies, they are 100% publicly owned by the local authorities and are managing public assets to give an essential public service. Moreover, the companies are run by Boards of Directors representing the various stakeholders involved. However, it is difficult for stakeholders in the Water Services Boards and Water Service Providers to hold their Directors and top managers accountable as they are not shareholders strictly speaking.

Old and Dilapidated Infrastructure

The water and sanitation companies have inherited old, dilapidated and in some cases obsolete infrastructure from the local authorities, the National Water Conservation and Pipeline Corporation and/or the Ministry of Water and Irrigation. The existing water supply networks (i.e. piping) have long passed their economic life which increases the unaccounted-for-water through frequent bursts and leakages. For example, some of the pipes in Eldoret date back to the 1920s. Furthermore, the water supply networks are serving more people than they were initially designed for. So far, rehabilitation of the existing infrastructure is yet to be fully achieved.

Old Staff in a New Outfit

All the water companies have inherited employees previously employed in the Department of Water and Sewerage of their respective local authorities or like in Homa Bay from the Ministry of Water and Irrigation. Only the directors and managers have so far been hired competitively. Besides having a cost implication this action brings into question the kind of employees the company has been forced to inherit. These employees, although in a new outfit, are likely to be slow or not ready to adapt to the reforms. For example, some employees have carried to the companies their previous corrupt practices and inefficiencies – denting the corporate image of the companies. The situation in Homa Bay is even more complicated. All the employees of South Nyanza Water and Sanitation Company are still being paid by the Ministry of Water and Irrigation and therefore answerable to their 'employer' rather than to the company. This makes the work of the Managing Director quite difficult because he has to manage a workforce not directly answerable to him.

Inadequate Capacity to Manage the Increasing Demand for Water

The populations in these municipalities have increased and will continue to do so. A major challenge to the water companies will be to provide enough water, in quantity and quality, to the increasing population. Presently, none of the companies has yet met the daily water demand of their respective municipalities. Apart from Nakuru Water and Sanitation Services Company which has so far met at least half of the residents' estimated demand for water (i.e. a supply of 35,000 m³ per day versus a daily demand of 70,000 m³), the situation in other towns is not very promising. That is, 18,000 m³ versus 45,000 m³ for Kisumu Water and Sewerage Company; 3,000 m³ versus 18,000 m³ for South Nyanza Water and Sanitation Company and 2,000 m³ versus 9,500 m³ for Gusii Water and Sanitation Company. This supply-demand shortfall has resulted in frequent water shortages and the now familiar water rationing.

Persistent Water Problems

Photo: Sam Owuor

In an effort to increase the production of water, the water companies have embarked on rehabilitating their water pumping stations and (old) distribution networks; constructing new intake pumping stations; installing new water pumps; and increasing the pumping hours. Most of these are being accomplished through donor support or donor-funded projects, i.e. the French government for Kisumu Water and Sewerage Company, the LVWATSAN programme for Homa Bay and Kisii, and the African Development Bank (ADB) in Nakuru. Through the ADB project, Nakuru Water and Sanitation Services Company is now able to supply water for 10 hours a day unlike the 6 hours it used to do. With the completion of the ADB-funded project, supply is projected to increase to 16 to 18 hours a day.

Limited Resources and High Costs of Operation and Maintenance

As much as the water companies are supposed to run as commercial enterprises, they are incurring very high operation and maintenance costs compared to the revenue they are collecting. This brings into question the economic viability of the companies. In addition, limited resources affect the achievement of the well-written and ambitious strategic plans, business plans, performance contracts and benchmarks, which have come to be synonymous with the water companies. Sometimes the companies are not able to achieve them at all. The recent increase in electricity tariffs in the country are a further burden to the water companies who rely largely and sometimes wholly on electricity to pump water not only from the treatment plants to the consumers but also from the water intake points to the treatment plants. A pumping scheme is much more expensive to run than a gravity scheme. In Kisumu, for example, the electricity cost of pumping water from the lake to the treatment plant is "very high". The implication has been a reduction in the hours of pumping water to consumers, and unreliability of the supply and rationing.

Local Political Interference

In its strategic plan (2007-2012), Kisumu Water and Sewerage Company points to "local political interference" as one of the risks towards good governance, financial resource mobilization, promotion of efficient utilization of resources and effective communication to stakeholders and customers[17]. It can be argued that political interference is bound to occur given the fact that the water companies are wholly owned by the local authorities which are by their very nature political. Politicians will always have a tendency of interfering with proposed water projects and appointment of Boards of Directors.

Inherited Debts, Liabilities and Multiple Fees

The water companies are paying too many fees, putting more pressure to their already constrained operation and maintenance costs. Of these, 10% of total revenue goes to the municipality as a fee for lease of assets, 5% is paid to the Water Services Board and another 1% is paid to the Water Services Regulatory Board. In addition, the water company is expected to pay 5 cents per every m^3 of water to the Water Resources Management Authority. For a period of time the Municipal Council of Kisumu exempted Kisumu Water and Sewerage Company from paying "dividends and rates" to the council until "they start making profit". Furthermore, the companies claim to have inherited debts and other liabilities from the previous service providers which they continue to repay or shoulder to-date.

Extension of Water Services to Low-income Neighbourhoods

Poor planning has made it difficult for municipalities to put up a water infrastructure especially in the mushrooming informal settlements. Their illegal status – at least according to the municipal authorities – has hindered the expansion of municipal services to serve them. The LVWATSAN programme in Homa Bay is tackling this problem by combining its water and sanitation interventions with town planning. Increasing access of water to the urban poor has also been hampered by the perception that the poor do not have the capacity to pay, yet they are paying far much more to get water from water vendors and other sources.

Lack of Autonomy for Major Investments

Since the companies do not own the assets, they are only allowed to do minor investments. Water Services Boards who own and manage the assets are the ones responsible for investment. For any new investment, the Water Service Providers have to get approval (through a sector investment plan) from the Water Services Boards, implying more bureaucracies that it was intended to reduce.

Inevitable Flat Rate Tariffs

Flat rate tariffs, especially in Homa Bay and Kisii, will continue to persist as long as the problem of lack of meters and malfunctioning meters is not addressed. For example, over three-quarters of the connections are on a flat rate tariff because most of the meters "stopped working long time ago", according to one of the respondents. In Kisii, a flat rate tariff is being used in areas where the likelihood of the meters being stolen is high, as meters are very expensive to replace.

Illegal Connections

Despite the efforts towards controlling, reducing and stopping this unsustainable habit of consumers, all the water companies are concerned about the persistence of illegal connections, not only in low income areas as one would expect, but also in other parts of the city. In Kisumu, for example, there are cases of illegal connections even in schools.

Conclusion

This chapter has outlined some of the emerging impact and challenges of water sector reforms in urban water supply. Whereas the emerging impact and challenges are in line with the results of other studies and findings in Kenya[18], there is need for a detailed research on some of these issues. One area that has so far received little attention is the impact of the water sector reforms and interventions on the livelihoods of the urban poor households. It has been observed that improved access to safe, reliable and affordable water, especially to the urban poor, can be beneficial in a number of ways, for example:

• It can improve the household's health and nutritional condition by reducing waterborne diseases and morbidity;

- It reduces the amount of time and energy spent on looking for or fetching water, which can be used on other productive economic activities;
- It reduces the high cost of buying water from other sources and the cost incurred in treating waterborne diseases, which can otherwise be put to other uses;
- It increases participation of women in income-generating activities and the girl-child's school attendance; and
- It can enhance the household's economic activities that depend on water (i.e. small-scale businesses, urban farming, etc).

Notes

1. Krhoda, G.O. 2008. *Building Broad Coalitions around Decentralized Institutions in Water Sector Reforms in Kenya*. Paper presented at the "NCCR North-South programme – JACS East Africa – CETRAD/DOGES scientific workshop", Naivasha, December 15-16, 2008.

2. Kisima. 2007. *Will SWAPs fix the Water Sector? Kisima Issue 4,* January 2007.

3. Mumma, A. 2005. *Kenya's New Water Law: An Analysis of the implications for the Rural Poor*. Paper presented at the international workshop on "African Water laws: Plural Legislative Frameworks for Rural Water Management in Africa", Johannesburg, January 26-28, 2005. Kisima. 2007. *Will SWAPs fix the Water Sector? Kisima Issue 4* January 2007. Gakuria, F. 2008. *Water Sector Reforms*. Guest lecture presented to the University of Nairobi, Department of Geography and Environmental Studies' students, Mombasa, December 9, 2008.

4. Kenya, Government of. 1986. *Sessional Paper No. 1 of 1986 on Economic Management for Renewed Growth*. Nairobi: [Government Printer].

5. Mumma, A. 2005. *Kenya's New Water Law: An Analysis of the implications for the Rural Poor*. Paper presented at the international workshop on "African water laws: Plural legislative frameworks for rural water management in Africa", Johannesburg, January 26-28, 2005. Ngigi, A. & Macharia, D. 2006. *Kenya: Water Sector Policy Overview Paper*. Paper presented to Intelligent Energy, European Commission.

6. Kenya, Government of. 1999. *Sessional Paper No. 1 of 1999 on National Policy on Water Resources Management and Development*. Nairobi. [Government Printer]. Gakuria, F. 2008. *Water Sector Reforms*. Guest lecture presented to the University of Nairobi, Department of Geography and Environmental Studies' students, Mombasa, December 9, 2008. Krhoda, G.O. 2008. *Building Broad Coalitions around Decentralized Institutions in Water Sector Reforms in Kenya*. Paper presented at the "NCCR North-South programme – JACS East Africa – CETRAD/DOGES scientific workshop", Naivasha, December 15-16, 2008.

7. Kenya, Government of. 2006b. *Kenya National Water Development Report: A Report Prepared for the UN World Water Development Report II*. Ministry of Water and Irrigation, Nairobi.

8. Kenya, Government of. Water Act 2002. Section 57(5)(d). Nairobi. [Government Printer].

9. Kenya, Government of. 2006a. *The National Water Resources Management Strategy (NWRMS), 2006-08*. Ministry of Water and Irrigation.

10. Ministry of Water and Irrigation (MWI). 2005. *Strategic Plan of the Ministry, 2005-09*.

11. Kenya, Government of. 2006b. *Kenya National Water Development Report: A Report Prepared for the UN World Water Development Report II*. Ministry of Water and Irrigation, Nairobi.

12. Mumma, A. 2005. *Kenya's New Water Law: An Analysis of the implications for the Rural Poor*. Paper presented at the international workshop on "African Water Laws: Plural legislative frameworks for rural water management in Africa", Johannesburg, January 26-28, 2005.

13. Ministry of Water and Irrigation (MWI). 2005. *Strategic Plan of the Ministry, 2005-09*, Nairobi.

14. Water Services Regulatory Board (WASREB). 2008. *Impact: A Performance Report of Kenya's Water Services Sub-sector – Issue No. 1.* Nairobi.

15. Kisima. 2008. *Water Sector Reforms: Five years on. Kisima Issue 4* May, 2008.

16. Kenya, Government of. 2006b. *Kenya National Water Development Report: A Report Prepared for the UN World Water Development Report II.* Ministry of Water and Irrigation, Nairobi.

17. KIWASCO. 2008. *Kisumu Water and Sewerage Company Limited Strategic Plan 2007-2012.* Kisumu.

18. Wambua. 2004. Ombogo. 2006. Otiego. 2006. Citizen's Report Card, 2007. Kisima, 2008 and WASREB, 2008.

SECTION TWO

Management and Practices

Urban Water

Strategies for Industrial Water Management in Kenya: Loopholes in the Existing Institutional Arrangement[1]

Joseph Onjala

Industrial Water Demand: The Challenge in Kenya

Water is an important input in both manufacturing and industrial processes. In the total use of public water in Kenya, industry is still a minor user, only consuming 4% of the public water supply. The relatively small industrial water use however masks substantial differences in the industrial water consumption and its impact on the water resources. In the urban areas the manufacturing industry consumes 13% of the public water and in the large urban centers the industrial consumption has been much higher i.e. in Nairobi (39%), Nakuru (26%), Kisumu (22%), Thika (23%), and Eldoret (17%)[2]. The overall demand structure for water resources in Kenya is given in Table 1. The recent water demand estimates show that industrial water demand is expected to grow from 220,000 cubic meters per day in 1990 to 366,000 m^3 per day by 2000 and 491,000 m^3 per day by 2010.

TABLE 1: WATER DEMAND, 1990–2010 (THOUSANDS OF M3/DAY)

Demand by Category	1990	2000	2010
Domestic water			
Urban	573	1,169	1,906
Rural	532	749	1,162
Industrial	219	378	494
Irrigation	3,965	7,810	11,655
Livestock	326	427	621
Inland fisheries	44	61	78
Wildlife	21	21	21
Total/Day	5,680	10,615	15,937
Total m3/Year (Millions)	2,073	3,874	5,817

Source: Kenya, Republic of 1992[3]

Kenya's 2nd National Water Master Plan showed that the urban water consumption was 35% in 1990 and was expected to rise to 45% in 2010. The urban daily use is highest (34.7%), followed by rural (32.2%), livestock (19.8%). Based on these projections, the proportion industrial water is expected change from 13% of urban water consumption in 1990 to only 12% in 2010.

The above figures, nevertheless, underestimate the impact of the industrial water use in Kenya, first because in addition to the public water many of the large industries consume large amounts of water from the rivers and private boreholes, i.e. about half of the industrial firms use water from rivers and own unmetered boreholes. Secondly many industries send polluted effluents into the rivers, thereby degrading water resources on which the population downstream is dependent. As Kenya continues with its quest for industrialization, it is expected that there will be increased demand for water to meet production requirements.

The quality of river water in Kenya is still considered to be generally good.[4] Thus the problem is that most development experts and institutions think that environmental pollution due to industrial activities in Kenya is insignificant. However, water resources are constantly under threat due to local pollution, particularly where there are intensive industrial, agricultural and human settlement activities. Although the total pollution load on the environment is still relatively low due to the low industrialization level, the social impact of water scarcity and industrial pollution is significant due to the direct dependency on open water resources by Kenyans. Furthermore, the proportion of the population without access to adequate water in Kenya is still very high, making it essential to conserve the available resources. River water is generally neutral to slightly alkaline with some rivers having slightly acidic headwaters. Emissions of organic pollutants from industrial activities are becoming a major cause of degradation of water quality in Kenya. Every year large volumes of water are extracted by industries and in return toxic effluents and hazardous waste are released into the riverine environment for the downstream users. A significant number of chemicals persist in the environment and cause widespread soil and water pollution. The downward migration of pollutants from the soil into the groundwater can especially be problematic in Kenya where groundwater is often directly used for drinking without any prior treatment.

The industrial sector in Kenya experienced rapid growth in the early and late 1960s with an annual growth rate of about 6%. Most of the industries are based within or near major urban centers. Such industries include tanneries, textile mills, breweries, creameries, paper production and recycling plants, chemical processing industries and slaughterhouses, which discharge effluents into the existing sewers and thus exert more pressure on the existing sewerage infrastructure. Water pollution arising from urban-based industries is evident in Nairobi (chemical processing, paper recycling, and slaughterhouses), Webuye (paper production), Thika (tanneries, textile industries, and chemical processing), Athi River (slaughterhouse and tannery), Nakuru (chemical and textile industries), Kisumu (fishing, chemical processing, and agro-processing), and Mombasa (fishing industry, chemical industries, oil industry, and agro-processing). In a recent incident, householders in Mombasa turned on their taps to find that the water had traces of oil in it. The contamination came from the leakage of an underground diesel storage tank into the drinking water supply. Increasingly, Kenya's agricultural produce from irrigated areas have been rejected from export markets because of contamination by heavy metals from industrial wastes.[5]

Some agro-processing industries are located in rural areas. They include coffee pulping and fermenting industries, tea processing, sisal fibre processing, sugar processing, and canneries. The discharges from these industries are mainly organic with high Bio-chemical Oxygen Demand (BOD) loads. For example, sugar milling consumes large quantities of water and the discharge has a BOD range of 3,000 to 5,000 mg/l.[6] The quality of the final effluent is invariably poor and depresses the oxygen levels of the receiving waters such as Nyando, Kuja, and Nzoia rivers. Indeed, data on emissions of organic water pollutants show that the load increased in Kenya from 26,834 kilograms per day in 1980 to 49,125 kilograms per day in 1998.[7] The industry shares of the emissions of organic water pollutants load in 1998 was distributed among several industrial sectors in Kenya as follows: Primary metals (4.1%); Paper and Pulp (12.2%); Chemicals (5.9%); Food and Beverages (68.4%); Stone, Ceramics and Glass (0.1%); Textiles (8.8%) and Wood (1.9%).[8] As a result of their strategic location upstream, the waste from industrial activity in Kenya is concentrated where it has the most severe impact.

Kenya is among the countries listed to run short of water in the next 25 years[9] (The other countries are Ethiopia, India, Nigeria, and Peru). *Population Reports* points to a number of culprits in the impending *water trap* that could warrant urgent policy and institutional measures:

• The primary issue is the rapid population growth and industrial development, coupled with the finite nature of the freshwater resources. The scale of severity is visible in the lack of water services in Kenya, with between 12 and 14 million people without access to safe water and over 20 million without adequate sanitation.

• Secondary explanations linked to the impending water scarcity are: climate variability, water catchment degradation, droughts, pollution of waterways, inefficiency of water use among industrial and domestic users, rapid urbanization (requiring higher per capita water use) and sub-optimal water allocation between different activities. The extent of pollution of water resources is already drastic, impacting most heavily on the majority of the population in rural and urban areas.

• In absolute terms, the demand for water was expected to increase by between 300 - 400% over the period of 20 years from 1990 to 2010.[10] The proportion of industrial consumption was also projected to decline as water recovery in Kenya was assumed to reach half of the industrialised countries' level by 2010, a feat that has been impossible in Kenya, given the lack of sound pricing and regulatory enforcement.

Given the current physical constraints in the water resource availability in Kenya, it is unlikely that the expected surge in water demand can be met. In view of the above challenge, water policy and institutional structures must be made appropriate to the emerging social and economic contexts in Kenya.

Industrial Water Management: Implications for the Institutional Arrangement

Industrial water management can be rational only if the institutions responsible for such management are efficient. Institutional arrangements establish the basis for *market* and *administrative control* over water use and wastewater disposal patterns. The theoretical literature elaborating the gains possible from institutional arrangement—both in the general and in the water sector contexts – are vast and growing.[11] The institutions create order and relative certainty for water users and facilitate achievement of economic and social goals. When not operating properly, institutions establish impediments to efficient resource use and significant resources must be expended (i.e. higher transaction costs) by individuals to compensate for their poor design.

Industrial water management in Kenya poses very special problems for institutional analysis. Externalities from wastewater pollution, common pool resources, i.e. ground and surface water abstractions, asymmetric information and exhaustibility are among the labels that can be applied to some of these special water issues in Kenya. Given the existing constraints to water resource availability in Kenya, the most significant implications of the appropriate institutional arrangement for industrial water use are as follows:

• Pricing of public water should discourage profligate use of industrial water in order to deal with exhaustibility (sustainability). This issue concerns the capacity of the water authorities to enforce pricing and regulation on the public water use to ensure efficiency. If the enforcement of water tariffs is weak, then the effectiveness of pricing as a policy instrument for industrial management may be limited severely. In most cases, the authorities in Kenya are still preoccupied with revenue raising rather than water use efficiency.12

• There should be an effective way of ensuring excludability in the use of public, surface and ground water sources (as opposed to public water) and a means for checking the open access character of the water so that water exploitation does not become a victim of the "tragedy of the commons". There is widespread conjunctive use of ground, surface (river) and public water sources by industries. Access to the three sources of water implies that any constraint in one source leads to shifts to other alternative sources since in most cases they are substitutes. This raises the question of excludability i.e. how successful are the water authorities in Kenya in controlling access to these captive sources? Lack of effective control renders the tariff increases for public water less effective in inducing water conservation behavior among industries.

• Enforcement of wastewater practices should allow efficient water using practices to become embedded in the industrial firms' technology. This can only be achieved through appropriate regulatory and incentive frameworks.

Existing Structures for Surface and Ground water Management in Kenya

The 2002 Water Act provides for the separation of the management of water resources from the supply of water and sewerage services and the separation of operation from regulation and policy making. This has been achieved through the operationalization and establishment of the following institutions:

• The *Water Resources Management Authority* (WRMA), which has the responsibility to manage, protect and conserve national water resources;

• Six *WRMA Catchment Offices*, to give the Authority a presence in different regions of the country, facilitating the policy of decentralized water resource management;

• Six *Catchment Area Advisory Committees* to advise on water conservation, use and allocation (including the issuance, cancellation, and variation of water permits at catchment level and associated fees);

• A number of *Water Resources Users Associations* to make possible community participation in the management and development of water resources and related environmental issues and to serve where necessary as fora for conflict resolution;

• The *Water Services Regulatory Board (WRSB)*, the overall regulator of water and sewerage service delivery;

• *Water Services Boards (WSBs)* licensed by the WRSB to be responsible for the efficient and economical provision of water services;

• A large number of *Water Service Providers (WSPs)* operating under Service Provision Agreements (SPAs) with WSBs to provide water and sanitation services to consumers within the areas of jurisdiction of the WSBs;

• The *Water Services Trust Fund (WSTF)* to finance the extension of water services to poor communities – deploying donations, grants and funds allocated by the exchequer to further the goal of achieving universal access for all Kenyans to water and sanitation;

• The *Water Appeals Board (WAB)*, an independent body established to resolve disputes between holders of water rights and any other dispute arising within the water sector which cannot be resolved at a lower level.

The Ministry of Water and Irrigation is charged with the responsibility for policy formulation, overall supervision and guidance in the sector and for sourcing funds for water resource management and development from the exchequer and development partners. The WRMA is the primary water resource manager. At the national level it acts as a Leviathan that ensures the viability of the resource countrywide, while the Catchment Offices serve as regional facilitators for decentralized governance structures. The Catchment Area Advisory Committees are served by a combination of local and national administrators and are intended to advise water users on the mechanisms for water management and regulations governing water use.

Water Resource Users Associations are community - level groups. They are intended to be the primary decision-makers. They determine water community water resource needs and communicate those needs to the Catchment Offices to obtain assistance in development of the structures necessary to meet those needs. The WSRB, operating parallel to and in close coordination with the WRMA, is the national regulatory authority in the water sector. It governs the Water Service Boards and the Water Service Providers operating at the local level. From this description it should be clear that the intent is to invert the flow of decision-making from the Ministry of Water down to the community. The idea is that communities make decisions about what they need and must only conform to the regulatory authorities to ensure the sustainability of the resource for them and for other communities sharing the resource.

Loopholes in Existing Institutional Arrangements

Uncertainty and High Transaction Costs
Efforts have been made to regulate industry surface and groundwater withdrawals through licensing restrictions for wells or through spacing norms to maintain ecological sustainability. However:
• The borehole completion records currently filed in the Groundwater Section of Ministry of Water and Irrigation (MWI) would be the basic data for future groundwater resources management. Of the nearly 10,000 boreholes in Kenya, only 11% have some water quality data logged on the MWI database. Most of the data represents those tested at the time of initial drilling, but no testing has been undertaken for most of the boreholes since then. The scarcity of existing data makes it quite difficult to meaningfully assess the groundwater potential and quality in the country.
• Abstraction point investigations are usually inadequate due to limited equipping of water bailiffs, leading to arbitrary apportionment. No enforcement of compliance with the granted permits by the water bailiffs due to inadequate staff and logistical support. There have been less funds to mobilise geologists, hydrologists and water bailiffs to undertake field monitoring exercises. At the same time, demand for river water resources has increased substantially (for domestic, livestock and irrigation). Additionally, access to the rivers has increased dramatically due to settlement/ subdivision of riparian land, meaning that the number of commercial abstractors has grown dramatically. The resulting absence of tight government control for ground and surface water exploitation has effectively provided an open access condition in which industrial water abstractors have abstracted on a "take as much as possible" basis.
• Current measures have sought only to regulate the establishment of groundwater structures, rather than the quantum of water extracted. This strategy overlooks the need for controlling profligate water use by industries.
• Issuance of surface and ground water permits has been mostly unpredictable, involving lengthy processes. This often frustrates applicants who are willing

to follow the right procedures and rewards those who provide inducements or facilitation to the water authorities.

The prevailing institutional framework can be seen to be ineffective in impacting industrial water management due to the following responses. The attitude of most of industrial firms is to regard permits as theirs and view enforcement officers or water authorities as infringing on their rights to harvest from their own investments. Several issues arise from the regulatory enforcement:

• Under the existing a priori appropriation doctrine, firms that acquire the permit also indirectly preclude the award of subsequent permits to other industrial firms in the same location;

• Older and more influential firms have benefited from permit allocation at the expense of younger firms.

• Even though there are avenues for transfer of water rights to new usage, i.e. from industrial to agricultural or domestic, there are no channels for transfer of permits between users. These restrictions preclude sharing permit allocations or transfer of water services between adjacent industries. Restrictions on transferability of water permits are restrictions on efficiency.[13] Transferable property rights in water have become a common prescription for improving efficiency in water allocation.[14]

• Monitoring and regulation of groundwater/surface water abstraction is very weak and nonexistent in most areas. Furthermore, the enforcement approaches are also criticized by industries because the system of licenses tends to breed corruption and inequity. Such an approach also assumes a highly simplistic, centralized approach to the water allocation problem, and ignores the possibility of localized management options by Local Authorities.

The resulting absence of tight control for ground and surface water exploitation has effectively provided an open access condition in which industrial water abstractors have abstracted on a "take as much as possible" basis.

Some intricacies remain in the linkage between pattern of land ownership, location, industrial activities and the exploitation of groundwater or surface water resources for which there is no clear institutional framework for management of water resources in Kenya. For example, a number of industrial firms (i.e. in Thika) opposed the enforcement of sewer charges on grounds that they discharge their effluents within their premises; hence they do not contravene any government regulations. The current institutional framework does not limit the extent to which an industrial firm can pollute surface or groundwater within its premises. Yet much of the untreated industrial wastewater ends up seeping underground.

Poor Tariff Strategies by Water Authorities

The water tariff structure by Water Authorities is mooted to deliberately encourage industrial water consumption. The obviously flawed incentive structure could be streamlined and the revenue base of the water authority enhanced by a proper regulatory mechanism for ground and surface water usage. For most of the urban water authorities, the sourcing of ground or surface water by industries also represents foregone revenue and there is no clear mechanism for dealing with this.

121

Some of the water companies (in Nairobi and Nakuru) see the need to impose a sewer charge for the groundwater since the wastewater is discharged in municipality sewers. In some urban areas where the shifts to groundwater and borehole are expected to impact heavily on the Local Authority revenues (i.e. Thika town) due to changes in water tariffs, there have been contractual arrangements between the firms and water authorities for the firms to pay a concessionary rate for water consumption every month.

In some urban areas where the shifts to groundwater and borehole was expected to impact heavily on the Local Authority revenues (i.e. Thika town) due to changes in water tariffs, there have been frantic efforts by Water Authorities to dissuade firms from such shifts. This was done by contractual arrangements between the industrial firms and water authorities for the firms to continue consuming an agreed amount of water in return for waivers on sewer charges. It is apparent that such an incentive structure deliberately encourages consumption of public water rather than ground or surface water by industries. Such a flawed incentive structure could be streamlined through a proper regulatory mechanism for ground and surface water. The effect of the prevailing ground/surface water enforcement is that they render water tariffs enforcement by Local Authorities to be defective for industrial water management.

There are other aspects of the water pricing institution by Water Authorities that make it inappropriate for industrial water management in Kenya:

• Water pricing in the urban areas is not implemented with the express goal of ensuring industrial water conservation, instead, the goal of water authorities is to raise revenue to support water service provision. All the local authorities place emphasis on revenue generation even though water service coverage remains low in all the urban areas. An appropriate pricing arrangement should be sensitive to conservation needs and the elimination of profligate use.

• Where municipalities are the water service providers, their ability to influence water consumption through tariff structure is very low. This is partly because the final responsibility for tariff approval lies with the Ministry of Local Government and not the Water Authorities.

• In all the urban areas water tariffs do not pose a credible constraint to industrial water use decisions. This is because the extent of liability enforcement for water consumption is quite limited. The existing laws or by-laws do not make it sufficiently compelling for firms to pay their water bills promptly while the liability rules remain vague.

Multiplicity Wastewater Standards for the Industries

There are two broad categories of wastewater standards for the industries in Kenya: (i) those stipulated by National Environment Management Authority (NEMA), usually applicable to discharges into streams and large water bodies; these standards are enforced by NEMA; (ii) Another set of standards are those stipulated by the Local Authorities by-Laws and enforced by the Local Authorities.

The second category of standards usually applies to firms discharging into public (municipality) sewer network. The two standards are related in so far as Local Authorities set the targets for the performance of their Sewer Treatment Works on the basis of NEMA discharge standards while they in turn stipulate in their by-laws, industrial wastewater standards compatible with capacities of their treatment facilities. The standards specify the limits for wastewater discharges based on a number of parameters that every wastewater disposing plant should take into account.

The main parameters include: Volume of the effluent; Bio-chemical Oxygen Demand (BOD); Chemical Oxygen Demand (COD); Toxicity: Suspended Solids (TSS); Permanganate Value (PV); Dissolved Solids (DS); Synthetic detergents/other chemicals among other parameters. The main problem with these multiple standards is that they complicate the regulatory enforcement in urban areas where firms have the options of discharging into both sewers and streams or rivers. This is because the jurisdiction for regulatory enforcement falls under different institutions. Most countries have multiple and extensive central government agencies in wastewater management. In Kenya, there are two main agencies in the enforcement structure; Local (municipality) Authorities and the National Environmental Management Authority (NEMA).

Local (Municipality) Authorities Enforcement

Even though all the Local Authorities stipulate the effluent discharge standards in their respective by-laws there are no mandatory requirements for the industries to undertake adequate measures such as the use of specified production technology before they dispose of their effluents. In industrial districts within urban areas, it remains impossible to monitor the quality of effluents discharged by individual firms especially if it is discharged directly into sewers. This weakness makes it ideal to impose "Trade Effluent Charges". However, trade effluent charges are also very difficult to implement and politically untenable. Only one town in Kenya, i.e. Eldoret town, has explicitly stated in its by-laws the intention to use this instrument.

The National Environment Management Authority (NEMA)

The NEMA has developed a procedure for vetting any new industrial firms. Proposals for industrial ventures that are presented to the Ministry of Industry are referred to the NEMA for clearance and matters related to wastewater disposal have to be addressed in the Environmental Impact Assessment (EIA). In order to evaluate the environmental suitability of firms, the Pollution Control Section requires information on: technical drawings of the factory layout including water supply and sewerage; detailed description of all the different industrial processes to be carried out; quantities of wastewater produced and the expected physico-chemical characteristics which may be obtained from simulated operations; a flow-scheme indicating the different streams of wastewater proposed pre-treatment/treatment process, storage, recycling, daily volumes and fluctuations, final discharge points and any other relevant information.

The above requirements mark the first step towards regulating the industrial wastewater in Kenya. The main weakness of this regulatory approach is that the requirements are placed at the planning stages, while certification of the wastewater treatment plants is required before industrial production can commence. Most industries commission their plants without completing their physical structures as laid out on the plant proposals. Industrial location policy particularly in relation to protection of rivers could become handy at this stage but often other matters of property rights (such as land ownership) predominate.

The Use of Enforcement Instruments by Local Authorities and NEMA

There are three (3) judicial steps in the enforcement of industrial wastewater standards in Kenya. The first step involves monitoring and inspection of the firms. The inspection could be triggered by citizen complaints, environmental meetings, self-reported samples or own field inspection by enforcement officers in NEMA or Local Authorities (LA). Court prosecutions represent the ultimate step in the enforcement process for wastewater regulation. We summarize the enforcement instruments in Table 2.

TABLE 2. SUMMARY OF ENFORCEMENT INSTRUMENTS IN KENYA

		Main strategies	NEMA			Local Authorities		
			Extent of Usage	Implementability	Effectiveness	Extent of Usage	Implementability	Effectiveness
Command and Control	Monitoring and Inspections	a)Self reporting b) Sampling	a) Moderate b) Low	a) Moderate b) Low	Low	Low	Moderate	Poor
	Warning Letters	Selective	High	High	Low	High	Low	Low
	Court Action	a) Fines b) Closures	a) Moderate b) Poor	Low	Moderate	Poor	Poor	Poor
	Water Permit	Cancellations	Low	Poor	High			
Economic Instruments	Financial Penalties	Violation penalties				Proposal in 2 towns	Poor	Could be effective
	Sewer Charge	% of water consumed				All towns	High	Moderate
	Trade Effluent Charge	Water volume				Proposal in 2 towns	Poor	Could be effective

Source: Onjala (2002)[15]

Problems with Multi-Agency Enforcement

There are several examples of impediments posed by multiple enforcement agencies in Kenya. Judgements during enforcement are influenced by the regulatory statutes for each enforcement agency. Some enforcement officers judge a firm to be non-compliant if in its officially analysed wastewater discharge, any of the parameters violates the stipulated concentration standards. Other officers are more lenient and they consider a number of parameters violated. Yet these judgments vary greatly between and within NEMA and the Local Authority. As a result of the multiplicity of judgements, firms shift allegiance between the Local Authorities and NEMA by changing their disposal points (i.e. between public sewers and rivers/streams) depending on which enforcement agency is perceived to offer better regulatory terms.

Towards a Sustainable Institutional Arrangement

Experience with water management calls for "decentralising water management to the lowest appropriate administrative level"[16]. One of the major flaws in Kenya's institutional arrangement for industrial water management has been an attempt by the Water Resources Management Authority (WRMA) and Local Authorities to simultaneously create capacity at the national level to deal with the water management problems at the local levels. This practice, however, negates a "*subsidiary principle*" in water resource management. As a rule, national governments should not implement tasks that can be done more efficiently or effectively at lower government levels, although they should ensure that these tasks are executed (subsidiary principle). There is need to restructure the roles and responsibilities of the WRMA, which could then transfer water management functions to service providers in order to foster institutional efficiency in the water sector.

Current theoretical models of enforcement also suggest that, other things being equal, a broad-based enforcement agency is likely to be able to achieve higher rates of wastewater compliance than a group of smaller ones. A wastewater regulatory agency that covers a wide geography and/or a variety of media is more likely to be able to identify and exploit synergies i.e. by implementing compliance-enhancing deals, rather than one with a narrower range of jurisdiction[17]. We consider that most of the problems encountered during enforcement of wastewater standards to emanate from a low level of implementation rather than any serious deficiencies in the legal provisions. We attribute the low level of enforcement to:

• The multiple agencies with different mandates, each agency trying to define the laws and activities on the basis of its narrow mandates;

• The division of responsibilities among multiple agencies, with little coordination, making it procedurally difficult to harmonize operations, thus leading to lack of interest among some departments that felt disadvantaged;

• A notable absence of what can best be described as the "philosophy of enforcement" (or internal policies and principles) by the various government departments – a shortfall that we attribute to apathy emanating from the multiplicity of agencies. This apathy is greater at senior levels where, to some extent, it is reflected by the comparatively junior grade officers responsible for enforcement while the senior officers engage in non-enforcement activities.

Notes

1. This study was conducted as part of the DANIDA/ENRECA activity organised jointly between the Institute for Development Studies, University of Nairobi and the Centre for Development Research, Copenhagen. I am grateful to Dr. Poul Ove Pedersen and Dr. Stefano Ponte for their useful comments.

2. Kenya, Government of. 1992. *The National Water Master Plan*. [Government Printer].

3. *The National Water Master Plan*.

4. Nyaoro, 2001.

5. Mogaka, Hezron., Gichere, Samuel., Davis, Richard & Hirji, Rafik. 2006. Climate Variability and Water Resource Degradation in Kenya: Improving Water Resources Development and Management. *World Bank Working Paper No. 69*. The World Bank.

6. Mwango, F.K. 2000. *Capacity Building in Integrated Water Resources Project in IGAD Region*. An IGAD publication.

7. World Development Indicators, 2001. World Bank. *World Development Report 2000/01: Attacking Poverty*. Washington, D.C. World Bank. "Aide-mémoire." *Draft. Integrated Water Resources Management Strategy: Identification Mission*, September 28, 2001.

8. *World Development Report 2000/01: Attacking Poverty*.

9. *Population Reports*, Vol. XXVI, No. 1, September 1998. John Hopkins University Press.

10. Kenya, Government of. 1992. *The National Water Master Plan*. [Government Printer]. The Water Act, Cap. 372 of the Constitution of Kenya.

11. Olson, Mancur. 1971. *The Logic of Collective Action: Public Goods and the Theory of Groups*. Cambridge, Massachusetts: Harvard University Press.

12. Olson, L.J and Knapp K.C. 1997. "Optimal Resource Allocation in an Overlapping Generations Economy." *Journal of Environmental Economics and Management, Vol.32*, No.3. pp.277-292. Bromley, D.W. 1989. "Institutional Change and Economic Efficiency" *Journal of Economic Issues, 23* (3). pp. 735-759. North, D.C. 1991. "Institutions". *Journal of Economic Perspectives 5* (1). pp.97-112.

13. Onjala, Joseph O. (2002, Ph.D dissertation). Managing Water Scarcity in Kenya: Industrial Response to Tariffs and Regulatory Enforcement. The Department of Environment, Technology and Social Studies, Roskilde University, Denmark, January 2002.

14. Colby, B.G. 1998. Regulation, Imperfect Markets, and Transaction Costs: The Elusive Quest for Efficiency in Water Allocation, pp. 488 in Bromley D.W. (ed). *The Handbook of Environmental Economics*. Blackwell, Oxford UK.

15. Colby. Regulation, Imperfect Markets, and Transaction Costs: The Elusive Quest for Efficiency in Water Allocation. pp. 485.

16. Onjala (2002, Ph.D dissertation).

17. Alcázar, L., L.C. Xu, &A.M. Zuluaga. 2000. "Institutions, Politics, and Contracts: The Attempt to Privatize the Water and Sanitation Utility of Lima, Peru".

18. Heyes, G.A. & N. Rickman.1999. Regulatory Dealing – Revisiting the Harrington Paradox. *Journal of Public Economics, 72* pp. 361-378.

Further Notes

- Daily Nation, January 28, 2001. Features. Available at: http:www.nationaudio.com/news/dailynation/28012001/index.html.

- Hurwicz, Leonid. 1998. Issues in the Design of Mechanisms and Institutions, in Loehman T. Edna & Kilgour D. Marc. 1998. *Designing Institutions for Environmental and Resource Management.* Edward Elgar Cheltenham, UK.

- Narain, V. 1998. "Towards a New Groundwater Institution for India". *Water Policy 1.* pp.357-365. Otieno, F.A.O. 1991a. *Pollution of Nairobi Rivers: A Research Report to Nairobi City Council and Office of the President.* Nairobi, Kenya.

- Otieno, F.A.O. 1991b. Chief Sources of Water Pollution in Kenya. *Journal of the Institute Management, Vol.10,* No.3. pp. 42-46, .

- Otieno, F.A.O. 1995. Role of Industries in Sustaining Water Quality: Sustainability of Water and Sanitation Systems. 21st WEDC Conference, Kampala, Uganda.

- Ruttan, V.W. 1998. Designing Institutions for Sustainability, in Loehman T. Edna and Kilgour D. Marc (eds.). *Designing Institutions for Environmental and Resource Management.* Edward Edgar , Cheltenham, UK.

- Saleth, R.M. & A. Dinar. 1999. "Water Challenge and Institutional Response: A Cross – Country Perspective ". Rural Development Department, The World Bank.

- World Bank, 2004. *Towards a Water-Secure Kenya: Water Resources Sector Memorandum.* Washington, D.C.

The Setbacks of Mismanaged Urbanisation: Pollution of Rivers in Nairobi

Mathieu Mérino

Introduction

Water resources in Nairobi are currently a critical issue. Supplies are limited in terms of quantity and quality, while demand increases due to population growth and economic development. Water shortages, sometimes called "city droughts", have been widespread since the beginning of the 1980s. At the time, the total city water supply was 116 million litres a day whereas the city required 156 million litres.[1] This shortfall has never been bridged. Though efforts have been undertaken, especially through the 1989 Nairobi Third Water Supply Project,[2] Nairobi is still short of water. At present, the city requires an average of 360 million litres a day.[3] The Nairobi water supply relies on four main sources, which when combined have the capacity to provide an average of 400 million litres a day. The Ngethu Dam and the Sasumua Dam, both located on Chania River, north of the city, provide more than 90% of Nairobi's water, with 300 and 60 million litres a day respectively.

Smaller amounts are supplied by Kikuyu Springs and the Riuru Reservoir, on the Athi river drainage basin, north-west of the city. Nevertheless, Nairobi faces a recurrent insufficient water supply. Since 1999, the city has been hit by heavy power and water shortages, the worst being in 1999–2000 when Nairobi residents faced several months of rationing. Seasonal factors are often pointed out as the main reason for city droughts. However, these only worsen an already mismanaged situation. The water distribution in Nairobi is caracterised by deterioration. Most of the networks supplying the city were established about three decades ago and have neither been improved nor expanded to cope with the increasing demands of a population of 3 million inhabitants.[4] According to United Nations Programme for Human Settlements (HABITAT),[5] 50% of Nairobi's water goes to waste because of leakages and other loopholes in the supply system. Recent improvement of water infrastructures has cut down loss of water to 37%.[6]

Nevertheless, the water supply issue is not the same throughout the whole city. Although water coverage in medium and high-income settlements is high, access to potable water remains a major problem in low-class and informal settlements.[7] According to a study by the World Bank,[8] 58% of Nairobi residents have no piped water. Within informal settlements, only about 12% of residential plots have running water while about 86% of the population obtains its water from kiosks.[9] Some estates have had no tap water since they were built, for example the rapidly-growing Ruai estate where there is no distribution network.

Consequently, these residents mainly rely on private vendors and they pay up to ten times above the legal rate.[10] They actually pay more for their water than high-income households with individual connections. Water at kiosks retails at Kenyan Shillings KSh 3–4 per 20-litre jerrycan, increasing to KSh 5–20[11] or more during shortages.[12] Most residents in such settlements can not afford to buy the amount of water they require. This inadequate formal supply - insufficient in quantity and excessive in price - leads people, especially slum dwellers, to obtain water from more accessible sources: from roof-catchments or open sewers and drains and also from the nearby rivers.

Nairobi lies within the Upper Athi River basin and is traversed by a number of rivers and streams, the main ones being Nairobi River, Motoine-Ngong River, Mathare River and Gitathuru River[13]. Historically, rivers in Nairobi have played a key role in the emergence and development of Nairobi as an urban centre. One of the reasons why Nairobi was chosen as a suitable stopping place by the railway builders in 1899, was the cool clean water. The Maasai refer to this as "*enkare nairobi*". The first camps were set up in the area now known as Kiambu. In order to be closer to water supply sources, the settlers moved down to the area around the present City Centre where the three main rivers: Nairobi, Motoine-Ngong and Mathare flow. These rivers also determined the areas of settlement according to different social groups during the early periods of the city's growth. The more affluent, mostly European, settled in the northern and western parts of the city where river-water was clean before it was polluted downstream.

Today, the use of the rivers as a supplementary source of water creates a predicament as the quality of the water has greatly deteriorated. River-water, which was once clean and potable, is now often stinking, heavily polluted, and dark-coloured. Nevertheless, river-water is still widely used by many people for drinking, washing cars as well as for disposing waste. This is done with total disregard for environmental and sanitary regulations. The use of such polluted waters is actually quite risky, especially for slum populations who are their first users. Ingestion of such water or simply contact with it in its stagnant state favours the development of various diseases, such as malaria, typhoid, bacterial and amoebic dysentries and giardiasis. The situation has become so severe that fauna and flora have disappeared from certain sections of the rivers. A case in point is the Nairobi Dam. Once a place of recreation, fishing and sailing, it is now infested with the water hyacinth. The whole Athi River area and its population are also threatened by pollution from the rivers in Nairobi. The rivers feed into Athi River, which is the main source of water for people downstream in the drier districts of Machakos, Kitui and Kilifi.

Nairobi's river pollution should be considered in view of the daily sanitary hazards that users encounter as well as the accompanying environmental deterioration. Added to this is the fact that the rivers are prone to flooding complicating the situation further. Pollution is partly caused by the extensive use of its waters and banks by the urban poor particularly low-class and informal settlements, characterised by scarce resources of land, municipal services or money. Rivers have become an integral part of numerous activities of slum inhabitants, small farmers and informal workers.

The first part of the article aims to show how the rivers interplay in the city's failed urbanisation. In 1999, the Government adopted a single-legislative approach through its much touted *Environmental Management and Coordination Act*. However, this was not enough to improve management. In actual fact, the legal protection of rivers requires a specific definition of the actors' function. As long as local authorities remain incapable of providing adequate services, it seems that the future of rivers is dependent on co-ordinated action by its first users, the urban poor. In relation to waste, for which insufficient public management has led to pollution of rivers, involvement of the adjacent population seems to be the only viable alternative that would improve the quality of both low-class and informal settlements and the rivers.

Pollution of Nairobi's Rivers: A Result of The City's Urbanisation Failures

Figure 1: Nairobi City

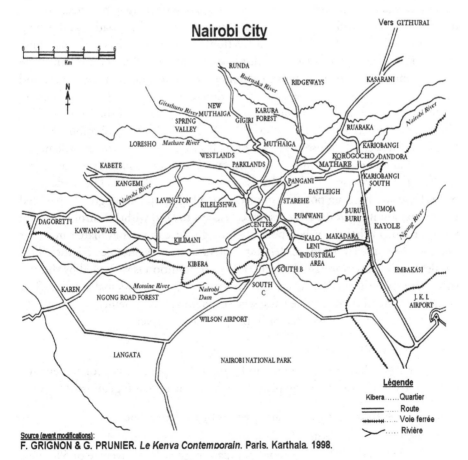

Source (avant modifications):
F. GRIGNON & G. PRUNIER. Le Kenya Contemporain. Paris. Karthala. 1998.

The three main rivers in Nairobi are, from north to south, Mathare, Nairobi and Motoine-Ngong rivers. Originating from the Kikuyu escarpment, the Mathare River first enters Karura Forest before flowing through Muthaiga, a high-class residential estate. After the bridge on Thika Road, the river flows through the Mathare Valley slums, the low-class Kariobangi North settlements and then through Korogocho slums, before joining the Nairobi River. The Nairobi River originates from Ondiri Swamp and enters the city through the Dagoretti settlements. The river then flows through Kawangware and Kangemi low-class settlements before entering Kileleshwa, a high-class residential estate. It then flows through the City Centre, before reaching Gikomba market, Eastlands, Kariobangi and Dandora estates. The Motoine-Ngong River originates from Riu Swamp, north-east of the Dagoretti forest, then flows through the Ngong Forest. Just after the horse-racecourse area, the Motoine flows through Kibera slums and empties into Nairobi Dam. The Motoine River leaves the Dam as the Ngong River, and passes through Industrial Area before its confluence with the Nairobi River near Njiru shopping centre.

The river system is heavily polluted with domestic, industrial and agricultural waste. Its waters and banks are disparately used with damaging effects on its quality and river profile.[14] Domestically, the water is used for bathing, washing, swimming, as well as for children's recreation, circumcision rites and sometimes for drinking when people cannot afford to buy water. Solid and human waste is also deposited here. In a more economic-oriented approach, the rivers and banks are used for watering livestock, irrigation, cultivation, car washing, soaking waste paper, construction and quarrying activities. Sections of the tributaries are said to be polluted 2,000 times above the World Health Organisation (WHO) standards,[15] especially those running through low-class or informal settlements, industrial areas or an agricultural unit. In fact, pollution, whatever its cause, is a result of the weaknesses of the urbanisation process in Nairobi. The inadequacy of land-use and housing policies coupled with a high population growth have led to over-exploitation of the river resources. Access to land within Nairobi is such a serious problem that the urban poor are forced to use the last available areas, for example along the rivers, for settlement, cultivation and economic activities. This has devastating environmental effects. The river pollution is worsened by the fact that these unauthorised areas are not covered by basic municipal services. The authorities' inability to provide basic services to low-class settlements as well as to the industrial area is a major factor in the pollution of rivers.

Extensive Use of River Valleys

The river banks have been invested in by the urban poor, either for settlements or for cultivation, thus flooding plains which are the remaining, affordable areas.

Land Pressure: Forced Settlements along Rivers by the Urban Poor

In Nairobi, the municipality has not succeeded in managing the high housing demand generated by the city's rapid population growth (an average of 5% since the beginning of the 1990s[16]). As a consequence, many informal settlements have developed. According to a study by African Medical and Research Foundation (AMREF)[17], 55% of Nairobi's population lives in slums. Due to the strong land pressure in the city, most informal settlements are located on flood plains or landslides, in abandoned quarries, on the steep banks of river valleys and on vacant land such as next to dumpsites. Nairobi's main slum areas - Mathare, Korogocho and Kibera, which are also the city's most densely populated areas, are located within flood plains along the rivers.[18] Some areas are even built on swampland that has been drained and filled. Large parts of Nairobi's city centre are an example of such land. Other swamps in the Zimmerman, Kahawa West, Riruta and Kawangware areas have been drained, giving way to high-density settlements.

The effects are devastating for rivers as an extension of the settlements increases erosion and siltation, both phenomena reducing the flow and absorption capacities of the river channels. In addition, these new constructions bring with them quarrying activities which have the same negative impact on the rivers. Generally speaking, the river valleys in Nairobi are not suited for these high-density settlements. Low-class and informal settlements have developed on the eastern and southern parts of the city where the soil is mainly the poor-draining black cotton soil. This characteristic, associated with the rivers' valley erosion, has negative effects on the longevity of construction, especially on cement. Siltation in rivers, domestic drains and storm water channels is worsened by garbage dumping. This increases the flooding risks in these areas that are built on flood plains and have poor-draining soils. As a consequence, during the rainy season, areas where houses are constructed close to the rivers experience serious floods, as seen in 1978, 1981, 1986, 1998 and 2000. The worst-hit victims in all cases were the slum inhabitants who live along the banks of the Nairobi River. During the 1998 El Nino rains, most residents of informal settlements lost property. The most affected areas were Mathare Valley, Kibera, Karura, Ruaraka, Gikomba and Pumwani. Parts of the industrial area, Nairobi West, the area around Nairobi Dam and low-lying areas around Githurai, Roysambu, Ruaraka and Nairobi-Thika highway were also affected.

Figure 2: Nairobi City Flood-Plain settlements

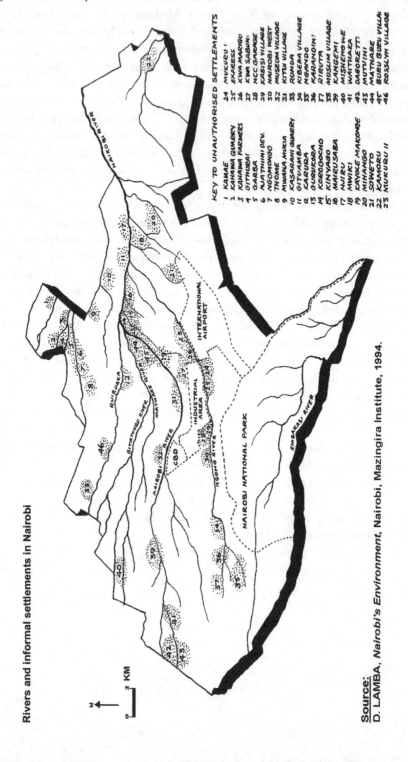

Rivers and informal settlements in Nairobi

KEY TO UNAUTHORISED SETTLEMENTS

1 KAMAE
2 KAHAWA QUARRY
3 KAHAWA FARMERS
4 GITHURAI
5 GARBA
6 NJA THINI DEV.
7 NGOMONGO
8 THOME
9 MWANA MUKIA
10 KASARANI QUARRY
11 GITUAMBA
12 KARUBA
13 GURUIGORA
14 KOROGOCHO
15 KINYAGO
16 MAIZU SABA
17 NJIRU
18 MWIKI
19 KAYOLE MAKONGE
20 MIHANGO
21 SOWETO
22 KAMORU
23 MUKURU II
24 MUKUKU I
25 EXPRESS
26 KWA MACHO
27 KWA SABUNI
28 NCC GARAGE
29 KABISI VILLAGE
30 NAIROBI WEST
31 MUSEUM VILLAGE
32 KITUI VILLAGE
33 RUNDA
34 KIBERA VILLAGE
35 NGANDO
36 KARANDINI
37 RIRUTA
38 MUSLIM VILLAGE
39 KANGEMI
40 MISHENGWE
41 WAITHAKA
42 DAGORETTI
43 MUTUINI
44 MATHARE
45 BURU GURU VILLA
46 ROSSLYN VILLAGE

2 KM

N

0

Source:
D. LAMBA, *Nairobi's Environment*, Nairobi, Mazingira Institute, 1994.

Development of Urban Agriculture

The capacity of the rivers is further weakened by the development of informal urban agriculture along the river channels. This urban agriculture is considered a major factor of pollution. The soil exposed for cultivation erodes and produces sediments that end up in rivers while agro-chemicals such as fertilizers, pesticides, and herbicides are washed away into the rivers. Additionally, farmers often use diverted sewage flow to irrigate their crops. The sewage treatment works in Ruai receive 80% of the expected loading, the rest having been diverted to irrigate vegetable gardens.[19] In Nairobi, about 20% of households practise urban farming, the majority of which are low-income households that produce subsistence crops.[20]

These gardens are located mainly near the central part of the city, often along roadsides and on flood plains and in high density residential areas in the eastern side of the city. An informal type of commercial agriculture has also developed along the river valleys. In the north-west and western parts of the city, tree nurseries and potted plants are common. These are within the high income areas where the products can be readily marketed. For instance, the leaveways of the Nairobi river, as it flows through Kileleshwa or Westlands along James Gichuru Road is not only highly cultivated with green plants, but also with vegetables, sugarcane, napier grass and bananas. Along the Mathare River, this informal commercial agriculture is seen in Spring Valley and in the Lavington-Muthaiga high-income residential areas. Cultivation aimed at the local markets is also developing along the rivers in Dagoretti where small farms have been established to supply the nearby Dagoretti market.

At Gikomba market, pillars supporting the traders' stalls are even erected inside the Nairobi River. All the vegetable waste is thrown into the river and the area around it used as latrines as there are no toilets close by. The real agricultural pollution however, is located upstream, originating from larger farms using chemicals. The north-western parts of the city upstream are extensively used for this type of agriculture. In these areas, the well-drained and fertile soil may produce various crops such as maize, beans, Irish potatoes on a large scale or a variety of vegetables, bananas, flowers and fodder. Consequently, rivers are heavily polluted when they flow into the city. For instance, Mathare River flows upstream through coffee and horticultural farms into the valley, where many chemicals are washed into the river during rainy seasons. Coffee factories also empty their waste into the river. Nairobi River is also extensively used by large-scale farmers, as in the Dagoretti area. Along the Motoine-Ngong River, for example in Karen, dairy cattle fodder crops and other high value horticultural crops are grown on large farms as vast as ten hectares.

The erosion of the banks and siltation of the channels due to their intensified use are clearly weakening the rivers' capacities, especially during the rainy season. Unchecked urbanisation has had further adverse effects on the rivers as the absence of basic infrastructure, such as waste management, has led to heavy sewage pollution.

Beyond River-water Pollution : The Inefficiency of Waste Management

The main source of river pollution is clearly identified as being wastewater. Wastewater includes sewage, water used for domestic purposes and commercial and industrial wastewater. After being treated, wastewater should discharge into the natural environment, in this case, the rivers in Nairobi. However, a large quantity of wastewater is not adequately treated, in addition to the fact that it does not even reach the sewer system. The wastewater system in Nairobi both for sanitation and industrial effluents is inefficient, and sometimes non-existent. Consequently, although total coliform in drinking water should not exceed 10 per 100 ml and fecal coliform should be 0[21] (according to the WHO standards), in Nairobi the coliform counts between 30 and 1,800 and fecal *E. Coli* is registered in all the rivers.[22] In dry seasons, the rivers, which have naturally have a low discharge, are mainly comprised of wastewater.

Lack of Sanitation and Waste Collection in Low-class Settlements

The current system for collecting, conveying and treating waste water is overloaded. The major problem is that wastewater does not reach the treatment plants. In low-class settlements, when the system exists, it is often deficient. Sewers are non-functional because of washed-away sections; steel sewer pipes are eroded; blockages in the sewer lines and manholes occur due to dumping of solid waste into open manholes, deliberate blocking by urban farmers for irrigation or the entry of stones, boulders and silt due to the erosion caused by the intensive use of the river banks. This results in a perennial overflowing of sewers. These deficiencies show that large amounts of sewage do not reach the sewage treatment plants, thus polluting the rivers. The situation is worse in unsewered areas[23] such as informal settlements.

According to HABITAT,[24] 94% of the inhabitants of informal settlements do not have access to adequate sanitation. In these areas, pit latrines are the major method of excreta disposal. Nearly 70% of households in Kawangware, Kibera and Mathare rely on pit latrines.[25] But these are far from being adequate in number.[26] They fill up quickly, and lack of disposal and purification services leads to latrine overflow into areas surrounding the pit. When it rains, the waste drains directly into the rivers. Pit latrines built along the river banks are emptied directly into the river. An alternative means of excreta disposal is the "wrap and throw" method, often referred to as "flying toilets".[27]

The condition of Motoine River when it empties into Nairobi Dam displays the raw sewage pollution linked to the waste water from unserviced low-class settlements. The Motoine River channel marks the southern boundary of Kibera. As the river flows along the slum, its pollution attains very high levels before flowing into the Nairobi Dam. At this point the dam is completely eutrophicated because of raw sewage pollution. This pollution is responsible for the water hyacinth infestation of the dam which was once a source of potable water and a venue for recreational activities. A similar situation is found at the Mathare River where the extremely high level of pollution in Mathare North is caused by raw sewage. This happens despite the dilution effect of input from the Gitathuru River which joins the Mathare River just before the latter enters Mathare North slum. Although Mathare is a sewered area, sections of the *Mathare River Sewerage Trunk*, commissioned in 1973 to serve Pangani, Mathare and Huruma estates, are always either blocked or washed away.

The inadequacy of solid waste collection and disposal completes the circle of the rivers' destruction. Currently, solid waste management is inadequately operated, apart from medium and high income settlements and industrial areas where the problem has been partially solved by private garbage collectors.[28] In low-class and informal settlements, there are neither reliable public collection services nor designated disposal sites and some of the inhabitants cannot afford the services of small-scale private collectors. As a result, solid waste is disposed of in streams, drainage channels or heaped by the roadside or on open ground. Dumping of solid waste is even more obvious at bridges and crossing points. In addition to blocking the existing sewage system, solid waste blocks the drainage channels' system, posing serious sanitary hazards. The drain water, which at this point is a mixture of domestic sullage and solid waste, then stagnates, creating breeding grounds for parasites, mosquitoes and flies. During the rains, these blocked channels overflow, causing damage and often flooding.

Industrial and Commercial Waste

Several businesses and industries have been established along river banks because their economic activities require proximity to water. Although most of the city's manufacturing activities are considered "light" industries,[29] they are sources of considerable pollution. Nevertheless, the heavy metals' pollution remains below critical limits, except for lead which is used in the plastics industry, in batteries and paints, and as an additive to gasoline.

The Ngong River is especially susceptible to this pollution. As it flows through the dam, the water is partially cleansed,[30] undergoing a natural purification. The river is barely polluted as it flows through the middle-income estates of Nairobi West and South C, which have adequate sewer and water supply systems. The situation changes at the Mater Misericordae Hospital bridge, which marks the beginning of the industrial area and a stretch of informal settlements along the river valley, for example, the Mukuru slums. From there and on downward, the river receives pollutants from not only domestic but also, and mostly, from industrial sources. Although most of these industries are sewered into the *Ngong River Trunk Sewer* (constructed according to a 1950 master plan), some of them dispose of waste directly into the river through storm water drains, because either the system is broken or due to their own laziness. By the time the river traverses the Outer Ring Road, it is already murky with floating oils, grease and solids, making the Ngong River the most polluted river in Nairobi.

The river continues to be subjected to increasing pollution as it flows through unserviced low-class and informal settlements (Doonholm, Soweto and Kayole estates). The Nairobi River also receives polluted industrial and commercial effluents as it passes through the City Centre. Buildings, *jua kali*[31] garages and craftsmen's sheds are constructed very close to the river, and all their waste is directed into it. The concentration of pollution increases dramatically at Kamukunji bridge, just after the City Centre, registering the highest values in the river profile.

At Kariobangi bridge, the industries and informal *jua kali* works associated with poor sanitation infrastructure increase the pollution level. An additional pollution concentration is recorded after picking up more effluents from cottage industries in the Dandora and Korogocho areas, peaking where the river flows close to Dandora dumpsite and the nearby Nairobi City Council sewage treatment works. On the Mathare River, industrial pollution remains at low levels, except at the Mathare slums' sections that can be attributed to source input from the small-scale cottage industry activities in the area.

Rivers have become a key resource especially for the urban poor who use the rivers to ease their difficult living conditions. Nevertheless, rivers, as well as their inhabitants, are now endangered by the pollution issue and the global deterioration of the river profile. Although legislative advances on the environment have been recorded in the last decade, the near absence of a pragmatic and concrete approach toward river management makes it an un-solved issue.

The Need for an Integrated Approach: Sustainable Environment and Stakeholder Participation

Despite the obvious deterioration of rivers, its management has not become a political issue. It remains absent from the leaders' preoccupations, as the legislation testifies. Nevertheless, if a legal framework on pollutant discharges into watercourses is required, especially to control the large amount of industrial effluents, the improvement of such a complex issue as the use of rivers will only be achieved by a global approach which would be environmentally-oriented and involve all the concerned actors.

Lack of a River Management Statutory Legal Framework

Protection of rivers has no statutory and complete legal basis. Measures related to rivers and to the authorities which are in charge may be numerous, but are not adapted to the development of a sustainable conservation policy.

Random Legislation on River Protection

No legal effort has been made to define a rivers' policy as they are considered to be non-priority resources, and few clear lines of management exist. In fact, legislation on river management has to be gleaned from a wide range of statutes.

It originates from the fact that the institutional and legal framework for resource management in Kenya is predominantly sectoral, with specific resources governed by specific Acts.[32] For instance, the use of agricultural land is governed by the Agriculture Act while water resources are overseen by the Water Act. Consequently, the Acts that are directly relevant to river management are as many as listed below:

Those related to health (Public Health Act, Malaria Prevention Act, Food, Drugs and Chemical Substances Act):

• Those related to agriculture and industry (Use of Poisonous Substances Act, Agriculture Act, Suppression of Noxious Weeds Act, Pest Control Products Act, Irrigation Act, Petroleum Act, Factories Act);

• Those related to the environment (Plant Protection Act, Trust Land Act, Land Control Act);

• Those related to water resources (Water Act, Lakes and Rivers Act, Fisheries Act), those related to the authorities in charge of the water resources (Local Government Act, Chief's Authority Act and all the Development Authorities Acts for Kerio Valley, Lake Basin, Tana and Athi Rivers or Ewaso Nyiro River Basin); and

• The Penal Code.

The legislation then has to be gleaned from statutes whose primary objectives do not include protection of rivers. When water resources are tackled through the different Acts, only two are clearly targeted: drinking water and fishing. For instance, the principal statute on water bodies, the Water Act, which is targeted at *"the conservation, apportionment and use of water resources"*,[33] obliges users to obtain a licence in all cases of proposed diversion, abstraction, obstruction, storage or use of water, with minor exceptions relating to domestic use. The Water Rules provide for pollutant discharges in a certain number of respects. Nevertheless, the rules do not really refer to rivers. Effluent discharges within a watercourse are recognised as an offence only if they would harm fish.

The Fisheries Act and the Lakes & Rivers Act are also limited. The former prohibits discharge of pollutants only if it causes harm to fishery resources, interferes with fishing or becomes a hazard for irrigation. The Lakes & Rivers Act limits its provision to the protection of animal life and navigability. The Penal Code also limits the offence to fouling of water from any public spring or reservoir.

The Public Health Act, that by far contains the widest range of provisions on polluting discharges, does not clearly refer to rivers. It has a duty to prevent and prosecute those responsible for any nuisances. However, it limits the polluting sources to drain and sewer discharge, and no standards of river water quality have been set.

This vagueness surrounding river management is explained by the long-standing lack of political interest in environmental concerns. It is worth noting, faced with the huge problem of potable water supply, the government only concentrates on protection of water bodies for drinking water. Thus, river water remains a secondary element. Additionally, the authorities do not wish to penalise the river users. Rivers are a necessity for the development of agricultural and industrial activities, and statutory and strict regulations on river pollution would jeopardise their interests. This then explains the low penalties. Under the Use of Poisonous Substances Act, the fine for abuse is KSh 1,000 while under the Public Health Act the penalty for failure to abate a nuisance as required by a medical officer of health is KSh 120 with a further daily fine of KSh 80 for continued failure to obey the order. Therefore, few industries are worried despite heavy pollution. The effluent from the paper and pulp mills of Webuye industries flows into the River Nzoia, as well as the waste from theNzoia and Mumias Sugar Refineries. This continues despite the fact that it has led to the annihilation of aquatic animal and plant life downstream on the Nzoia River.

In its instability, this legislative dissipation of river management also leads to enforcement problems. The policy on rivers empowers so many authorities and institutions that no coherent and effective policy can emerge.

Enforcement Difficulties

Protection legislation depends on different Acts, and as each Act generally includes the establishment of an institution in charge of implementation, there is a myriad of enforcing authorities and institutions mandated to deal with rivers.

• *The Water Act*: the Water Resources Authority is associated with a Water Apportionment Board and Regional Water Committees for each province;

• *The Public Health Act*: the Central Board of Health is the advisory body for the Minister while the enforcing institutions are designed as the local authorities; and

• *Other acts dealing with rivers:* They also empower the Agricultural Office, the Fish Warden, the Pest Control Products Boards, the Chiefs, etc.

As each institution or authority has a different mandate and is often linked to different water bodies, it is difficult to have a coherent policy. Some of the boards are single-issue oriented and may not necessarily take into account some of the interfering factors that may fall outside their mandate such as polluting discharges. Moreover, the water bodies are essentially managed by district and provincial-based institutions whether they are local authorities or provincial boards. In the case of rivers, administrative boundaries often do not bear any relationship to catchment areas. Most local authorities are also technically and financially incapacitated to fulfill the functions demanded by the different Acts. These include: providing an adequate sewerage and drainage system, giving licenses for and controlling effluents and punishing offenders. Moreover, up until 1999, local authorities enjoyed legal immunity from liability for offences therefore, enforcement of the Acts was not encouraged.

In situations where the environment is considered as a collection of single resources managed by different Acts and authorities, issues and goals of sustainable development cannot be addressed. This is especially true in this case where water and land issues are closely linked. Natural resources exist as part of complex, dynamic ecosystems and cannot be separated into static, uni-linear sectors, as until recently seen in Kenya. There were up to seventy-seven statutes that relate to the environment, thus fully justifying an integrated approach.

Towards an Integrated Approach

An efficient river protection system cannot be achieved without tackling different issues including water supply, waste management, agriculture land-use and methods of industrial development. Sustainable river management can only be attained if implemented from an integrated approach by taking all the involved resources, sectors and actors into account. This approach has been developed by the government through the Environmental Management and Coordination Act 1999. But the real issue concerns the environment's management actors.

Environmental Management and Coordination Act, 1999

The Environmental Management and Coordination Act of 1999 has been lauded for bringing to an end the sectoral resource approach paving way for a wider environment approach, integrating water, air and land. The environment is considered in its entirety through two main institutions: first, a National Environmental Management Authority, whose main function is to co-ordinate environmental management activities and second, the National Environment Council *"which shall be responsible for policy formulation and direction"*[34] has also been established. In order to show priority interest in environmental issues, the government has imposed statutory requirements, known as the Environmental Impact Assessment (EIA), a procedure for prior assessment of the environmental impact of a proposed development project.

The National Management Environmental Authority is charged with overseeing this process and issuing licenses. The Act also stipulates a mechanism for the dissemination of environmental quality standards. Until 1999, there were hardly any statutory environmental standards in Kenya. The country relied on World Health Organisation guidelines for water, which are health and not environment-based. A Standards & Enforcement Review Committee was therefore established to recommend standards for air, water and waste management, and then gazetted by the Minister of Health. Most importantly, the government has enhanced strong measures that criminalise environmental offences and that are supposed to remove impunity. However, it grants local authorities immunity from liability for nuisance caused by their incapacity to fulfill their responsibility in establishing and maintaining sewerage and drainage works. The Act also sets up a National Environment Tribunal coupled with high penalties for private and public actors. Polluting watercourses, for example, results in a penalty of up to 18 months imprisonment or a fine of KSh 350,000.

Environmental Management

The 1999 new legislation on the environment may be viewed as a considerable milestone but, on the other hand, it does not seem to solve the problem of enforcement. In Kenya, most statutes on the environment have aimed at snaring the populace. For instance, the Physical Planning Act was effected in 1996 in the aftermath of the furor against the government following the de-gazzettement of Karura Forest. The 1999 Environmental Act was enacted after several scandals about forestry resources and due to strong pressure from organisations demanding enforcement of the EIA. This new law is mainly aimed at appeasing public opinion rather than dealing with environmental issues. Even though the new legislation is credited with introducing the necessary policy coordination institutions, the implementation organs remain the same without reinforced capacities. Pressure has been especially exerted towards local authorities without questioning their capacity. Therefore, the new institutions seem to complicate an already muddled institution. The new legislation is also repressive, it does not deal with the issue of pollution prevention. As far as rivers are concerned, the new Act will have little impact. If it is enabled to act on the large industrial and agricultural pollution, it will not deal with the heart of the matter, which is the relationship between river pollution and the urban poor.

The Water Act 2002 suffers from similar weaknesses. The new legal framework involves the competences' allocation review and the creation of new institutions. Water resources management authorities and water services board at regional level have been recently established. The rationale is to devolve water management and service responsability from the Governement to these semi-autonomous bodies and local communities. The Ministry will retain a co-ordination and policy formulation role. The final phase which is now in progress involves the transfer of both utilities and personnel to the new bodies. The Ministry of Water and Irrigation has already handed over all water supply utilities to the new institutions. The Act also involves the creation of autonomous companies dedicated to water and sanitation provision, such as the Nairobi Water Company created in 2004. Neverthelss, the Water Act's core reform has still to be implemented. Indeed, the Act involves the definition of the National Water Resources Management Strategy. The strategy defines how water resources will be managed, protected, used, developed, conserved and controlled. The strategy for 2006-2008 has been prepared but has not been yet gazetted[35]. The new Act has also created a strong competition between public actors, who are now confronted with a loss of power. For instance, on the basis of the new Water Act, car-wash businesses located near the rivers were banned by the Ministry of Water in January 2003;[36] one month later, the Municipality authorised their resumption. This is a sector representing a major source of employement in Nairobi.[37]

In actual fact, the legal protection of rivers requires a specific definition of the actors' function. As long as local authorities remain incapable of providing adequate services, it seems that the future of rivers relies on co-ordinated action by its first users, the urban poor. In relation to waste, for which insufficient public management has bred pollution of rivers, involvement of the adjacent population seems to be the only viable alternative that would improve the quality of both low-class and informal settlements and the rivers. Among the poor, both groups and individuals are increasingly involved in waste management, especially solid waste. Three main solid waste management modes have thus been developed: re-utilisation, recycling and composting. Sorting of articles for re-use and recycling is done by the urban poor who look for re-usable items or those that can be sold. These include glass, bones, paper, batteries, organic waste, plastics and metal, among others. Young people and street children are specifically involved in this informal sorting.

The products are then sold to recycling industries through intermediaries. Several community initiatives are developing on composting. Besides the reduction in quantity of organic waste, composting enables small-hold farmers to produce an affordable fertilizer. This popular involvement has led to the development of an entire parallel economy centred on garbage. Besides partly solving the pollution and services' problems in informal settlements as waste is re-used, community participation in waste management is also an alternative to urban poverty. Sorting and composting activities offer potential earnings to an increasing population. These initiatives, often supported by Non-Governmental Organisations, have also developed in the fields of water supply and waste water management, leading to the launch of different environment awareness campaigns. Nevertheless, these stakeholder management modes suffer from lack of both encouragement and guidelines from local authorities.

Nevertheless, one encouraging institutional initiative has to be underlined: the *Nairobi River Basin Programme*, initiated by the United Nations Environment Programme (UNEP) in 1999. The programme, which brings together the UNEP, the municipality, several NGO's and residents, aims at restoring to the capital its riverine system as a resource of clean water and promoting a healthier environment for the people of Nairobi. Following the data collection's first phase, the second phase is involving six pilot projects implemented along the Ngong River and in Kibera. At Kibera, the project is currently building toilets with sewers connections, water-supply points, primary drainage and a small-scale recycling centre for the demonstration of good sanitation. On the other side of Kibera, at the Nairobi Dam, an inovative way of dealing with the invasive water hyacinth is underway (to harvest the hyacinth so as to manufacture marketable products such as paper and ropes). The third phase will deal with garbage collection. A pilot phase is going to involve the rehabilitation of a 2.5 kms' stretch from the Museum Hill roundabout to the Racecourse Road bridge (refuse removal and repair of sewers); the estimated cost, financed by the Kenyan authorities, is Ksh 150 million.[38]

Conclusion

Pollution of the rivers in Nairobi is among the most noticeable consequences of glitches in urbanisation. Inadequate land-use and housing policies have led to an excessive use of the rivers in a situation of high urban poverty. Erosion, siltation and pollution, whether from sewage or industrial and agricultural effluents seriously endanger the future of the rivers and of the slums built in the river valleys. The environmental legislative achievements of 1999 do not solve the issue. Even if the environmental statutes have been toughened, the central problem remains, which is the incapacity of the local authorities to cope with their basic liabilities. For this reason, the future of rivers in Nairobi lies in stakeholder participation in management of services and more specifically in a partnership between the population and the public authorities who can provide the necessary guidance.

Notes

1. *Daily Nation,* 3 February, 1982. 'Emergency water measures needed'.

2. The 1989 Nairobi Water Supply Project has added major new supplies from Thika Dam and Chania River which were expected to provide enough water to meet the estimated demand by the year 2005.

3. *Sunday Standard,* 18 May, 2003. 'More water problems for Nairobi'.

4. Kenya's 1999 Population and Housing Census. [Government Printers], recorded the population of Nairobi at about 2,143,000 people. Nevertheless, according to several studies, these statistics seem under-estimated. Some researchers assess Nairobi's population to be roughly 3 million: Davinder Lamba & Diana Lee-Smith, *Good Governance and Urban Development in Nairobi.* Mazingira Institute, Nairobi, 1998.

5. *East African Standard,* 30 July, 2002. 'The poor paying dearly for water-Habitat'.

6. *Standard Business,* 30 April, 2007. 'Nairobi water firm revenue hits Sh 3.3b'.

7. Bousquet, Anne. 2006. *'L'eau et les Pauvres à Nairobi: De L'apartheid Hydrique à la Fragentation Urbaine ; l'exemple de Kibera',* in Hélène Charton & Deyssi Rodriguez-Torres (eds.). *Nairobi Contemporain : Les Paradoxes d'une Ville Fragmentée.* Paris: Karthala.

8. *East African Standard,* 7 November, 2002. 'Kenya has water deficit'.

9. Okoth, P. F. & Otieno, P. 2000. *Pollution Assessment Report of the Nairobi Rivers Basin. Africa Water Network,* UNEP, Nairobi.

10. *Sunday Standard,* 15 April, 2007. 'Water vendors cash in on residents misery'.

11. Farid, M. 1999. *Small-scale Independent Providers of Water and Sanitation to Urban Poor: A Case of Nairobi.* Water and Sanitation Program & International Water and Sanitation Centre, Nairobi.

12. *East African Standard,* 6 May, 2003. 'Panic as water crisis hits Nairobi'. In May 2003, a jerrycan of water retailed at KSh 35-60 in areas where the demand was high such as Umoja, Buru Buru, Embakasi, Donholm and Greenfields estates.

13. See the map of Nairobi City, Figure 1.

14. Kahara, Sharon. 2000. *Nairobi River Basin Phase II : Pollution Monitoring.* NETWAS & UNEP Regional Office for Africa, Nairobi.

15. *Sunday Nation*, 28 January, 2001. 'Rivers choke in waste'. Two University of Nairobi lecturers, Dr D.O. Olaga & Dr N. Opiyo-Aketch, revealed: *"Unpolluted waters usually have Biochemical Oxygen Demand (BOD) – the measure for the amount of oxygen demanded by a water mass – values of 2 mg/litre oxygen or less. BOD for the Nairobi rivers have concentrations between 40 and 4400 mg/litre oxygen"*.

16. *Population Census*, 1989 & 1999.

17. *East African Standard*, 7 January, 2001. '55 percent of Nairobi population lives in slum areas-AMREF'.

18. See the map of Nairobi flood-plain settlements, Figure 2.

19. Obudho, Robert. 1999. *Environment and development in Kenya: Urbanisation and Management of Kenya Urban Centres in the 21ª century*. Nairobi: Kenya National Academy of Sciences.

20. Obudho. *Environment and development in Kenya: Urbanisation and Management of Kenya Urban Centres in the 21ª century*.

21. Total coliform bacteria are a collection of microorganisms that live in the intestines of man and animals. A specific sub-group of this collection is the fecal coliform bacteria, the most common member being *E-Coli*. The presence of fecal coliform in aquatic environment indicates that the water has been contaminated with the fecal material. The source water may have been contaminated by pathogens or disease producing bacteria or viruses which can also exist in fecal material. Some waterborne pathogenic diseases include typhoid fever, viral and bacterial gastroenteritis and Hepatitis A.

22. Okoth, P. F. & Otieno, P. 2000. *Pollution Assessment of the Nairobi Rivers Basin*. Nairobi: UNEP.

23. In Nairobi, only 10% of the population is served by a conventional sewer system and 20% by septic tanks. The majority of Nairobi residents i.e. 70 % only rely on manually cleaned latrines: HABITAT, *City Level Statistics about Water and Sanitation*, 2003.

24. HABITAT, 2003.

25. Lamba, Davinder. 1994. *Nairobi's Environment: A Review of Conditions and Issues*. Nairobi: Mazingira Institute; NETWAS International, "Water and Sanitation for Africa's Urban Poor", *Water and Sanitation Update, 13*, 2, September-December 2006.

26. HABITAT, 2003. In Kibera, there are often up to 200 persons per pit latrine.

27. Residents relieve themselves in their shacks and dispose the waste the following day. The exreta, wrapped in polythene paper, is thrown into open spaces where it is washed away into the rivers.

28. Japanese International Cooperation Agency (JICA).1998. *The Study on Solid Waste Management in Nairobi City in the Republic of Kenya*.

29. Light industries in Nairobi are activities such as agricultural processing, manufacture of textiles and other related products, repair works, motor assembly, leather tanning, manufacture of pharmaceutical and chemical products, paints and varnishes, distillers, bottlers, garages, printers and further downstream, abattoirs.

30. Although the water hyacinth has become a nuisance to the Nairobi Dam, it partly cleanses the water. The lowest pollution data recorded within the dam outlet station indicates that the dam, together with the water hyacinth, act as a sink for suspended sediments and organic matter and coliform.

31. *Jua kali*, a Kiswahili expression, means 'hot sun'. Generally speaking, it refers to the informal sector. More specifically, it refers to people who informally collect and recycle a wide range of materials so as to sell them.

32. U.N.E.P., *Environmental Law in Kenya*. Nairobi: Acts Press, 2001.

33. The Kenya Water Act, 2002, Section 5.

34. Wamukoya, George. 2000. *A guide to the Environmental Management and Coordination Act 1999*. Nairobi: Centre for Research and Education on Environmental Laws.

35. NETWAS International, "Water sector reform in Africa: Experiences from Kenya, Uganda, Tanzania and Zambia", *Water and Sanitation Update*, *13*, 1, May-August 2006.

36. *East African Standard*, 16 January, 2003. 'Karua bans vehicle washing near rivers'.

37. *East African Standard*, 5 February, 2003. 'NCC to issue licences to car washers'.

38. *Daily Nation*, 14 November, 2007. 'The murky Nairobi River could become clean again'.

The Potential of Grass-roots Leadership in Water Management in Fringe Communities: The Case of Dar es Salaam, Tanzania

Alphonce G. Kyessi

Introduction

With rapid urbanisation in poverty, the provision of infrastructure and services is not given priority by most homebuilders in the fringe areas of most urban centres including Dar es Salaam.[1] As time goes by, subdivision of plots and development of houses continues without accompanying infrastructure. The cost of extending basic infrastructure to the rapid growing fringe settlements of Dar es Salaam seems prohibitive and unlikely to be met by the Dar es Salaam City Council (DCC) and/or Dar es Salaam Water and Sanitation Authority (DAWASA) due to financial and management constraints.[2]

Availability of adequate water is a critical issue for meeting drinking, cooking and personal hygiene needs. Existing data on the incidence of water-borne diseases indicate that they are mostly prevalent where people use contaminated water or have little water available for daily use. Thus, considering that the public water supply system under the management of DAWASA does not reach most fringe areas, and that much productive time may be wasted by settlers, especially women and children, fetching water, it becomes necessary to identify available opportunities and coping strategies for providing potable water to fringe communities.

Sub-ward (*Mtaa*) leadership in informal settlements located in the fringe areas often organises settlers in the form of Community-Based Organisations (CBOs) that can assist in infrastructure provision. A number of options were sought by the residents in order to secure some basic infrastructure such as water supply and settlement accessibility. To explore the potential and the processes involved to access potable water in the fringe informal settlements, two settlements namely, Yombo Dovya and Tungi both located in Temeke Municipality, Dar es Salaam City were studied in 1999. Like in other inner city informal settlements, there have been several attempts by DCC, the *Mtaa* leadership, party organisations, private individuals as well as youth and women groups and the donor community to provide the basic services and facilities in Yombo Dovya and Tungi. This chapter discusses the possibilities and constraints existing in the provision of basic infrastructure to fringe settlements taking water management as an example. Potable water was chosen to explain this case as these settlements are not connected to the DAWASA water supply system and are remotely located. The purpose is to inform policy makers, researchers and practitioners including water providers and managers and water users on possibilities and constraints existing in the provision and management of water supply in remote and poor communities.

Potable water as an essential need plays a major role in health development and if water is not easily accessible, much time is wasted searching for it. One of the questions to be answered is how the residents, most of them being poor, are coping with the deficiency of water supply in these fringe areas. The other is, what techniques are being adopted, the main actors and their roles, how settlers are contributing towards the water improvement in the areas and what challenges or constraints exist in the areas. In particular, focus is on socio-spatial aspects, the water management process, community participation and self-help options, appropriate technology, links created, sustainability and poverty related issues in the provision process.

The Outskirts of Dar es Salaam: Focus on Two Informal Settlements

The profile of two settlements forms an important benchmark towards the discussion of basic infrastructure provision, specifically the water management in fringe areas. The issues being researched on include location, status of the neighbourhoods and local organisation. The two informal settlements selected for the fieldwork are *Tungi* and *Yombo Dovya*, in the outskirts of Dar es Salaam.

Tungi is located about five kilometres from the city centre along the Indian Ocean in Kigamboni Ward. It is a settlement which exhibits rural characteristics and has experienced fast growth especially after 1980.[3] By 1999 it covered 84 hectares of land and about 1,680 houses accommodating 17,500 people equivalent to a density of 20 houses per hectare and 208 people per hectare.

Yombo Dovya is situated in Yombo Vituka Ward, about 10 kilometres from the city centre of Dar es Salaam. The overall character depicts a rural environment with predominantly agricultural plants such as coconut palm trees, cashewnuts and mango trees. The settlement covers approximately 93 hectares and was first settled in the early 1900s.[4] It started to experience fast growth after 1980. By 1999 there were about 2010 houses equivalent to 22 houses per hectare with an estimated population of 24,000 and a density of about 258 persons per hectare.

The housing and population densities in the two settlements are considered very low compared to those of inner informal settlements which have densities of more than forty houses and about 500 persons per hectare respectively.[5]

Poor Neighbourhoods

A household interview conducted in both settlements indicated that most immigrants had shifted from both formal and informal settlements of Dar es Salaam searching for better opportunities. Citing declining employment opportunities in the traditional urban sectors, namely, service and manufacturing, and decreasing real income of most urban households, Kombe[6] asserted that the increasing occupation of land in the urban peripheries is a phenomenon that is clearly related to the changing socio-economic base. Engagement in urban agriculture has gained ascendancy in recent years as a survival strategy in response to the economic hardships of many urban wage earners to supplement declining real wages.[7]

The two settlements seem to exhibit a similar household occupancy pattern. The number of households per house in the settlements is about 2.0 and 2.2 with a household size of 5.2 and 5.3 in Tungi and Yombo Dovya respectively, indicating that a house owner shares a house with one tenant household. The findings show that in the fringe areas tenancy is a source of income for house owners. Further analysis of the data shows that tenants live in 48% and 57% of the houses in Tungi and Yombo Dovya respectively, most sharing a house with the house owner. House owners live in about 97% and 91% of the houses in Tungi and Yombo Dovya respectively, depicting one of the characteristics of informal settlements where a family builds a house mostly for its own accommodation.

Both settlements experience isolationism and a similar development pattern. The cost of commuting to and from work prohibits those employed in the city centre to look for accommodation in these two fringe settlements. Tenancy is still low compared with the old saturated informal settlements near the city centre, which accommodate a higher percentage of tenants of about 70% or more. In these old informal settlements, land for urban agriculture is not available and thus tenancy becomes the main source of income for the house owners. For example, in the case of Buguruni Mnyamani, tenants live in 77% of the houses mostly shared with house owners. It is also a common phenomenon that tenants would prefer to live in areas with infrastructure services (within or nearby) and employment opportunities. As observed in the field, the fringe settlements offer less of both opportunities unlike older informal settlements in the city.[8]

The average monthly total expenditure per household was TShs. 86,200 ($0.79 per capita a day) in Yombo Dovya and TShs. 99,600 ($0.89 per capita a day) in Tungi. These figures are higher than those found in inner city informal settlement of Hanna Nassif and Buguruni Mnyamani.[9] However, they also give a per capita expenditure of less than a dollar a day (about TShs. 700/=) implying that these residents are below the poverty line.[10] About 89% and 74% of the respondents in Yombo Dovya and Tungi respectively were spending less than US$ 1 a day.

Further analysis shows that 51% and 48% of the households in Yombo Dovya and Tungi respectively, spend less than 0.75 cents of a dollar implying that half of the population in the area is within the absolute poverty group.[11] Most (90%) of the households spend more than 50% of their income on food. Like in the inner informal settlements, the findings from this case show that the lower the monthly total household expenditure the higher the percentage of expenditure on food, leaving a very small portion of the income for other household basic services. The data suggests the need for innovative ways and means of providing affordable basic infrastructure for the poor households.

However, it appears that DCC does not have the capacity to collect property tax and service these fringe settlements. This suggests that if the *Mtaa* leadership is given the mandate to organise collection of property tax on behalf of the City Council they might be able to do so because of their local knowledge of the people and the area. Part of the collected property tax could be retained in the settlement, for example to create a revolving fund for improvement of basic services. However, one would doubt the capacity of the *Mtaa* leadership unless training on management of tax collection was offered.

Haphazard Development

Observations made from the two settlements indicate that habitation superseded the designation of land for installing infrastructure and the provision of community services and facilities (water supply, roads, markets etc.), contrary to the normative urban land management practice that envisages land servicing before habitation.[12] Like other inner informal settlements of Buguruni, Tabata and Hanna Nassif, these fringe areas have Town Planning Drawings (TP)[13] showing land use subdivisions prepared by the DCC and approved by the Ministry of Lands and Human settlements Development (MLHSD). However, Dar es Salaam City authority did not seem to have influenced a planned land development or enforced land development control in the two areas. Subdivision and sale of land was therefore going on haphazardly. This might lead to higher housing densification without setting aside land for basic community infrastructure facilities. Although most of the respondents (65%) showed willingness to survey their plots mainly in order to have more security over their properties, only 1% of the plots in both settlements has been surveyed. Surveying procedure of individual plots, which takes an unnecessarily long time, is done upon the owner's request depending on whether an owner has funds for doing so.

Basic infrastructure in the two informal settlements was inadequate or not available. Infrastructure facilities and services such as water, roads, nursery schools, health facilities and market are provided informally or spontaneously. All households used pit latrines. There was no drainage system except the natural drains. Solid waste collection was also lacking. Residents relied upon dispensaries and hospitals in adjacent settlements that are located about two kilometres away. The services were developed commensurate with demand as population grew. For example, roads were informally provided and were found only where landowners have hired trucks to bring building materials to their sites. Two primary schools, one in each settlement, were overcrowded and not sufficient to cater for the existing population.

The initial school at Yombo Dovya started with four rooms of four classes, built by the community and more classrooms are being added to meet the increasing number of pupils. Markets in both settlements were open-air stalls, initially established by households from their houses and gradually shifted to more "central or road junction areas". In spite of the informality in siting and provision of the facilities and services, most of them were fairly easily accessible to the majority of the residents in the settlements. However, none of these community infrastructures had land reserved for expansion commensurate with the present or future demands of the population. In some occasions, private individuals locate some of the services in their private plots, such as water kiosks or wells for community use, in order to ameliorate the scarcity of land for infrastructure. The rest of the chapter discusses the processes involved in the water provision and management in the fringe settlements.

Local Organisation

Local governance is increasingly becoming an important aspect of infrastructure provision and management in urban areas including fringe settlements. The establishments of environmentally sustainable and friendly planning and development systems are believed to be viable only if local communities play a central and strategic role.[14] Thus, institution building is essential in realising the outputs of the processes. Building local institutions and organisation geared towards infrastructure provision is commensurate with understanding the possibilities and constraints within the framework in which *Mtaa*[15] operates and the way the power structures (and relations) are managed for the day-to-day work at the local community level. Their is no doubt that the role and functions of *Mtaa* cannot be taken up by civil associations or the private sector. However, partnerships that may be created between them are considered to be vital in local economic and social development. As the local governments are legally recognised political entities, the organisation of *Mtaa* as the lowest unit of its administration to provide not only public services but also to maximise local participation and partnerships is vital to community development and poverty reduction.

The *Mtaa* Organisation Structure and Management

The management of the two fringe settlements, Yombo Dovya and Tungi, fall under the umbrella of the City Administration. Each of these settlements is one *Mtaa*.[16] The population and spatial size of *Mtaa* is not specified in the existing local government instruments and thus varies from one settlement to another. *Mtaa* as a local government administrative unit plays the role of co-ordinating community activities at the lowest local government administration level. *Mtaa* residents elect seven of its members as their leaders: Chairman, Secretary and five advisors. The *Mtaa* leaders administer the community more widely than Ten Cell leaders. The *Mtaa* serves, first, as a political/administrative link between residents and the upper level government structures that include the city and state. Secondly, the role of the *Mtaa* administration is to provide overall management of social and economic development activities for its people.

To enable the *Mtaa* leadership discharge their duties and functions without depending too much on technical and financial assistance from the upper central organs, and at the same time co-ordinate provision of basic community services such as potable water, solid waste collection, etc. a better working institutional structure is required. Because that structure does not exist, or has not been provided for in the existing legislation (except for two committees formed according to their sectoral policies of Education and Water), the two communities had put in place informal structures to work with the sub-ward leaders elected by the residents.

Bridging the wide gap created between the legitimate structure and the functions, the *Mtaa* Chairmen of the two settlements formed informal sub-committees to work with to enable them address the local community's problems. The committees comprised five to ten members and had specific roles. Most of the members are Ten Cell Leaders (TCL). The sub-committees include: *Executive Committee, Water Management Committee (WMC), Construction Committee, Health and Environmental Committee, Conflict Resolution Committee, Education Committee, Business Committee, Credit Committee*, and *Markets Committee.*

These committees are very important community development structures because they were composed of community members assigned specific responsibilities. However, some were active, others not so depending on the nature of activities being carried out in the community. The mode of operation was self-help (volunteering), which seemed to prevent some members from participating actively in the committees. In-depth discussions with the *Mtaa* Chairmen revealed that the committees work closely with Ten Cell leaders who assist in disseminating information to residents and resolving household conflicts. In total there were 40 and 46 Ten Cell leaders in Tungi and Yombo Dovya respectively. Each of the Ten Cell leaders headed about 40 houses. It has to be noted here that the CBOs operating in the old informal settlements such as Tabata Development Fund and Hanna Nassif Community Development Association are also utilising the Ten Cell leaders as a resource for better co-ordination and implementation of planned activities.

It appeared that the two settlements had a decentralised administrative and organisation system that resembles that of a registered CBO. Despite the limited economic ability by the *Mtaa,* they had been able to organise and manage all the committees and delivered good service to the people. Recognition of these structures by the DCC and fiscal decentralisation in the city administration structure ought to give the *Mtaa* some abilities/economic power of implementing community action plans prepared by the various committees. In many other settlements Ten Cell Leadership or local government did not have this strong legitimacy.

In Yombo Dovya the *Mtaa* leadership seemed to be active in performing its activities as envisaged by the community, the situation was the same in Tungi. Two thirds of the respondents were satisfied with the *Mtaa* performance. The level of satisfaction is a very important indicator in measuring performance of an organisation or institution. Based on the high level of satisfaction of the community there is no doubt that the *Mtaa* leadership has been fulfilling its mission of leading its community despite the problems raised by a number of respondents. Observations made in the settlements show that there was a sufficient leadership ability to work together with the residents in providing basic needs of the communities. In general, the *Mtaa* leadership had been playing a vital role in mobilising resources and organising community participation and private involvement in water provision to the fringe areas. The following section explains the methods used.

Collective Action in Fringe Water Provision Process

Participation in water supply and management in the two study settlements may be explained as a process whereby consumers influence the level or quality of water service provision to them. Participation in water supply to the informal settlements was a result of joint action between the public, the private sector and the community. Through their participation various resources such as expertise and experience, ideas, ability to organise, materials, labour and finance were mobilised and used to improve potable water in the two settlements. Various grassroots and external actors were involved. The grassroots actors include the water users i.e. individuals (households), vendors and the *Mtaa* leaders. External actors came in as advisors, financiers and contractors in the construction of the wells. The external actors include DCC, DAWASA, the Water Resource's Institute (WRI), the UNDP and Lions Club. While DCC has been acting as a facilitator, DAWASA staff have been providing free professional support in the form of, for example, chemical materials for treating water[17] to the Water Management Committees and to individual private owners. However, often technicians from DAWASA demanded some token pay for service, such as repair of the pumps provided to private individuals[18]. The Lions Club donated funds for the construction of 5 water wells in Tungi. After the construction the wells were handed over to the community. The Water Resource Institute provided technical support on location and construction of the shallow wells.

Community Involvement

The communities were involved in different capacities in the provision process of potable water. They participated in three different groups namely *Private vendors* (who have constructed own storage tanks and installed a pump for later use), *Water Management Committee* and *individual households* who have constructed their own wells for their own use.

Private Vendors

There are numerous private vendors in the two settlements, both small and big ones. The small ones, who have private traditional shallow water wells and sell water to other households, were 27 in Tungi and 14 in Yombo Dovya. The large vendors have constructed water storage tanks within their plots and pump water from deep wells dug on the plot or shallow wells dug along River Yombo. One of the private vendors in Yombo Dovya has constructed one shallow well with a distribution line connecting to three water storage tanks. In both settlements private vendors supply most of the water requirements for domestic use from wells. Three types of wells are common: traditional shallow wells, hand and motor pump shallow wells and motor deep wells. The vendors sell water at TShs. 20/= per 20 litres.

It is important therefore to note that a potential in private participation in water supply in the fringe settlements exists and it is sustained through the selling of water to residents where part of the revenue is used to maintain the wells. There is also free entry in water selling and this brings in competition of many suppliers. This process assists to regulate the price of water per litre and at the same time ensuring good quality water. This in a way might be the reason why the *Mtaa* leadership does not involve itself in regulating the price as some respondents complained. The emerging model of this private sector participation in the water provision in the fringe settlements has evolved through a *step-by-step improvement*. The vendors have been responding to demand of better service by building deep wells instead of shallow wells, which have salty water.

Water Management Committee

Residents, in collaboration with Ten Cell leaders, established the Water Management Committee to supervise water supply in the settlement. The leaders took the initiative of identifying and negotiating for space for locating shallow wells in their areas.[19] This is identified as community-based water supply. The approach adopted in the provision process took different forms in the two areas.

In Yombo Dovya, before 1994, the Community had contributed funds[20] through the initiation and co-ordination by the then village government (known also as Chama Cha Mapinduzi leadership)[21] of the area. Thereafter, local *fundis* (informal small contractors) living in the area who had skills in the construction of wells, were contracted by the village government to construct 12 traditional wells. In-depth discussion with elderly persons in the area revealed that the wells were not covered because of lack of enough capital funds. Water from these wells was fetched free of charge and thus as time went on the wells deteriorated and were later abandoned. Subsequently, households reverted to supplies from Yombo River and/or from vendors. Later on again, through an external assistance from the UNDP (through its programme of Local Initiative Facility for the Environment - LIFE) five wells were constructed and handed over to the community leaders (the Ten Cell leaders - TCLs) to manage and maintain.

Without a mechanism put in place to sustain the water improvement initiative, especially in operation and maintenance, it naturally collapsed. The case also shows that even with external assistance, if there is no cost recovery mechanism for operation and maintenance or if the system is not linked to other sources of revenue for operation and maintenance, the system will collapse. Suffice to say, a collective action must have clear rules, regulations and commitments embedded in it in order to sustain it. Commitments ought to include meeting cost obligations for maintaining common property such as the water supply. In Tungi, five hand pump shallow wells were constructed, in the late 1980s, with financial assistance from the Lions Club of Dar es Salaam and technical assistance from the Water Resource Institute of Dar es Salaam. The wells were handed over to the respective Ten Cell leaders. Water was also supplied free of charge in the beginning but later the community, through its leadership, in early 1990s resolved to introduce a water charge in order to maintain the pumps and the wells. Following the multi-party election at the *Mtaa* level in 1994, a Water Management Committee was formed as a committee responsible for water resources in the area and management and operation of the pumps. With advice from the Water Management Committee, the community privatised the management of the wells. Out of five public wells, three have been privatised. The operators do not own the wells but have a contract to run them on an agreed monthly fee.

Water Management in the Neighbourhoods

Various human, financial, natural and man-made resources were mobilised and were utilised in the water provision and management in the two settlements. These resources were mobilised from the community, private individuals (vendors), government institutions, training institution, NGOs e.g. Lions Club, local informal and formal businesses and the United Nations Development Programme (UNDP). Several resource mobilising strategies were adopted for the water supply in neighbourhoods without water from the DAWASA pipes as discussed above.

Underground Water: Source of Water

There is no DAWASA water supply system to either Yombo Dovya or Tungi. Many households in the two fringe settlements use traditional shallow wells within their plots for domestic water needs. In Tungi alone there are 29 wells. The owners normally maintain these wells but most of them are not covered posing the risk of possible water contamination. The general feeling of the households towards this source is that the water is salty and not safe. It ought to be noted also that unless housing densities can be regulated this source will be further contaminated due to continued use of pit latrines in the areas. As densities increase, a better option will be required.

All residents interviewed indicated that they obtained their supplies from wells. The average daily water consumption per household in Yombo Dovya and Tungi is more or less the same; that is 204 litres per day (or 38.5 litres per capita per day) and 206 litres per day (or 40 litres per capita per day) respectively.[22] According to the respondents, monthly household expenditure on water supply in both settlements ranges from TShs. 900 to 13,500 with an average of TShs. 4,310 or 4.3% of the total expenditure. The figures are similar to those found in Tabata where most people use water from deep wells.[23]

Meeting the Costs of Alternative Water Schemes

To meet water demand, the water schemes in the two settlements had mainly been financed through a process of self-help. Most households have built private wells at the household level using personal savings. With voluntary labour, one shallow well costs about TShs. 60,000 for materials. Wells built this way have been improved on a step-by-step process starting with traditional open shallow wells, then installing a hand pump, then upgrading them to deep wells with motor pumps and raised water storage tanks. Vendors had started to establish mini-distribution networks especially in Yombo Dovya with a purpose of having a wide catchment area. Some of the vendors e.g. those who had installed motor water pumps and storage tanks got financial support from relatives and friends, others sold farm produce.

It was noted that the vendor had continuously improved upon the mode of water provision on a step-by step process by using the income accrued from water sales. The adoption of better technology was prompted by the increase in demand of both quantity and quality of water. Funds for community-managed wells were obtained from within the community and external sources including Lions Club and UNDP. Lions Club provided about TShs. 75,000 in 1984 to DCC to construct 5 wells in Tungi. The wells were constructed by the Water Resources Institute in Dar es Salaam and handed over to the community for its operation and maintenance. UNDP through the programme of Local Initiative Facility for the Environment (LIFE) provided US $12,300 to DCC who contracted DAWASA to construct 5 hand pump wells in Yombo Dovya. After the construction they handed over the wells to the community for operation and maintenance. The sustainability of these wells is discussed later.

Property tax was not collected from the settlements during the period of the survey. This could have formed an important source of revenue to finance water supply, operation and maintenance and other community activities in the settlements instead of relying on external finance. Collection of this tax was centralised by DCC who had no capacity to organise its collection. It would have been a good idea to decentralise the collection of property tax to the settlement (*Mtaa*) level as now practised in the newly established municipalities. Building capacity at the local level is however necessary step for financial management.

Supplementing Household Incomes Using Revenue Accrued from Water Sales

Revenue from water sales had been used for operation and maintenance and to upgrade the systems. Water was sold at different prices depending on the source. Water from deep wells was sold at TShs. 20 per 20 litres. Water from shallow wells, which are covered and with hand pumps, was sold at TShs. 10 per 20 litres and water from shallow wells, which are not covered, was sold at TShs. 5/= per 20 litres. However, while water has been recognised by the community as a commodity, the Water Management Committee (WMC) of Yombo Dovya continued to provide water free of charge, a process that made the wells collapse.

Experience of the vendors shows that water supply system in the two settlements can be self-financing. It was noted that most of the revenue obtained from sales of water had been used to supplement household incomes, upgrade the water systems, pay salaries to water sales attendants, pay for operation and maintenance of the pumps and wells and pay for money borrowed. Replication or upgrading to a better technology had largely depended on cost recovery mechanisms.

Adoption of Appropriate Technology: Management of Wells

Technology is a system of organising and of taking action, which ultimately determines the way in which and the means with which objectives for the use of a technique are being reached. It involves the optimal combination of the factors of production (i.e. labour, materials, finance, organisation and information). It also depends upon the effective demand, the available resources and the affordability levels.

Labour for constructing the wells and operation and maintenance was contributed by the households, hired from the community and/or hired from outside the settlements such as the technicians from DAWASA who repaired water pumps. *Materials* and necessary equipment were purchased from shops or hired from DAWASA and other institutions. *Land* for installing the wells was donated by the residents. *Organisation* of the water supply is one of the important input resources. Individuals at plot or household level have done management of the water supply and other resources. In other situations Water Management Committees and other *Mtaa* leaders had the responsibility of managing the water supply in the settlements. However, in-depth discussions with various vendors and individual households revealed that the Water Management Committee (WMC) was not often efficient in water supply activities.

Informal discussions that were held with Ten Cell leaders in the settlements revealed that the WMC in Tungi had sub-contracted the day-to-day operation services of the community wells while that in Yombo Dovya was managing the community wells alone. Although the WMCs are supposed to be responsible for the water supply in general, this was not the case in both settlements. It is important to observe that without proper management of the water utilisation and use, sustainability of the underground water resource may be threatened. Mobilising the locally available community resources and constructing a piped system within the settlements with communal standpipes could form a sustainable option that may be considered and might result in better exploitation and use of the underground water resource. However, that should not be taken as the only option considering that the communities in the two fringe informal settlements have mixed income groups that have varying needs.

Adoption of Low-cost Technology

The technology applied, in the two areas, follows a demand-oriented service delivery and a community-based approach that could be rated as a low-cost system. In the areas, three main options were observed: traditional shallow wells, hand pump shallow wells and motor pump deep wells. These different sources met the requirements of the poor community with varying income expenditure in the settlements. The technology and standards used have evolved on a step-by-step improvement model termed appropriate technology that is easy to build, operate and maintain. It includes a small-scale type using a bucket to scoop water from an open well; medium-scale type using a hand pump and large-scale water supply using a motor to pump water to a raised tank (Table 1).

TABLE 1: PROCESS OF ADOPTING APPROPRIATE TECHNOLOGY
AND STANDARDS

Time Horizon	Step-By-Step Technological Change			Remarks
Stages	Stage 1	Stage 2	Stage 3	
Type	Traditional well	Hand pump wells	Motor pump and Raised tank with stand pipe	Technical support is sought when moving higher to another stage and possible with cost recovery
Operation and maintenance	Household members	Household members Hired sales attendant	Household members Hired sales attendant	Done on self-help Cost recovery also applied
Standard and scale	Small and open Shallow well Use a bucket to scoop water	Medium scale Shallow well Pump direct to a bucket	Large scale up to 9,000 cubic litres Water pumped from shallow and deep wells Use of gravity	Varying size depending on financial ability of the owner
Quality of water	Poor Salty	Poor Salty	Good from deep well Poor and salty from shallow well	Often quality tested by government laboratory

Source: Field study, 1999

Factors determining the need to change from one stage to the next were mainly availability of finance and the demand for better water quality.

Using Multi-model Water Management in Meeting Needs of the Poor

To meet the communities' effective demand, a multi-model system of water supply exists in the settlements with varying underlying participatory mechanisms. For instance, in Tungi supply is a partnership between the Community and private operators; private operators themselves and individual households; whereas as seen in Yombo Dovya, supply is by the Community Water Management Committee, private operators and individual households. The multi-model supply system existing meets the requirements of all income groups in the areas. The findings resemble those from a study on reducing urban poverty by Anzorena *et al.*[24] who caution that fixed models should be avoided in meeting the needs of the residents in a settlement because of different income groups[25].

Model 1: Private Management and Operated Water System

The three types of technology are used in this model. Some vendors/owners started with traditional wells then moved to hand pumps and gradually acquired motors that could pump water from deep sources or from wells located some distance from the residences. The individual household suppliers have constructed this system through their own finances. Normally, a member of the household sells water. The quality of water depends on the source and maintenance of the system. The price of water ranges between TShs. 5 and 20 per 20 litres jerry can. Water from shallow wells is sold for TShs. 5 and 10 per 20 litres jerry can while water from deep wells is sold between TShs 10 and 20 per 20 litres jerry can.

Model 2: Community-based Water Management System

The technology used is the hand pump from shallow wells. In Yombo Dovya, the households fetch water from the community wells free of charge. This model is built upon socialist related principles with centrally planned economies whereby citizens should get basic infrastructure services free of charge. Central organs were entrusted with the planning, distribution and maintenance of infrastructure using funds from the central government. The reliance on government led to collapse of four out of five wells constructed because of lack of revenue. In Tungi, on the contrary, households who fetch water from community wells pay TShs. 10 per 20 litres jerry can making most of the wells still operational. The wells are maintained using revenue accrued from cost recovery unlike the ones of Yombo Dovya. Residents in Tungi are getting water from these wells at a lower charge than from vendors. However, the water quality was noted to be poor and salty.

Model 3: Household Water Management System

Several individual households have own traditional wells within their plot. There are about 40 wells in Tungi alone. It cannot be concluded that these wells belong only to the poorer of the poor households because rich households own some. In-depth informal discussions with the residents of both areas, revealed that some households use water from this source for domestic purposes while other households supplement water from this type of well with safe water bought from vendors. The water quality is poor and salty. In conclusion, the residents are getting water from these three models, which supply domestic water to all the different income levels in the two settlements. Developing new techniques is dependent on the improvement of household incomes in the two areas, resource availability and the local capacity to manage them.

The maintenance system in place in Yombo Dovya and Tungi depended on the mode and type of water provision system in place. *Private management and operated water system* had a very elaborate maintenance programme whereby each owner cleans his/her well(s) and storage tanks and makes repairs as the need arises. Observation from the settlements shows that vendors with well-maintained and clean wells sell more. It was decided by the community that whenever a pump, that belongs to the *Community-based Water Management System*, is broken, people using it have to contribute money for its repair. However, it has been difficult to get sufficient funds for maintaining the wells because water was/has been provided free of charge. Over time, some of the pumps and wells have started deteriorating. Presently, four pumps out of five are out of order because some of the wells located away from residences (along Yombo River) were vandalised and parts stolen during the night. This system has collapsed because of the absence of an effective community-based operation and maintenance programme. In Tungi water was also being supplied free of charge in the beginning but later the community (in early 1990s) resolved to charge in order to get revenue for maintaining the pumps and wells. Ten Cell leaders under the party (CCM) were entrusted with the wells and thus sold water to obtain funds for operation and maintenance.

In Tungi, the CCM branch was advised to privatise the management of the wells by Mr. Mundo, the Instructor at the Water Resource Institute who was later elected Chairman of the Water Management Committee formed in 1994. Out of five public wells, three were privatised; therefore the operators did not own the wells but had a contract to run them on an agreed monthly fee.[26] However, it was decided that revenue from one of the three-privatised wells be used for *Mtaa* Chairman activities.[27] The Chairman was contracted to manage it. Free water supply was seen to be unsustainable. The Chairman of the WMC explained that the idea of selling water and later privatising some of the wells took long to be accepted. The need to privatise was however, necessary because all the hand pumps were broken and the residents could not raise funds needed for repair. Presently, the pumps are working and the community is getting cheaper water from this model than from the private vendors.

For *Household Water Management System,* household members normally undertake operation and maintenance. One may wish to speculate that, with increasing densification these traditional wells are threatened with faecal contamination from pit latrines, which is the only sanitation system available in the two areas. Without development control in the area, continued use of traditional wells threaten the health of the people and thus the urgent need for alternatives to an affordable water system for the communities.

Self-help in Water Management

Self-help means people endeavouring to achieve goals through their own efforts. It is an individual or collective response to situations perceived to be unsatisfactory, which people seek to overcome by sustainably improving their living conditions and increasing their self-reliance.[28] Close to 60% of the respondents indicated that they had been involved in self-help activities in the two settlements. The residents had earlier voluntarily participated in road maintenance, water supply, solid waste collection and school building. In fact, the people themselves had initiated most of the water wells projects found in the two areas. More importantly, about 90% of the respondents further said that they would be willing to work in partnership with the government to improve the piped water supply in their area. The respondents indicated that they would be ready to freely contribute their labour. This evidence shows that residents have an interest in improving their living conditions.

Most settlers were willing to form community-based organisations because they believed that through these, association residents could be enlightened more, sensitised and that togetherness would bring development faster than individualism. These respondents believed that *Mtaa* or Sub-ward is a government unit where people cannot make decisions contrary to the wishes of the government. They think that there is control from above. Certain procedures and rules have to be adhered to. As has been discussed, the two communities are operating as CBOs. They however, do not have constitutions and they are not registered as CBOs. As observed, there seems to be similar functions of the committees established and the way they operate in the two settlements and that of active CBOs in inner city settlements such as the Hanna Nassif Community Development Association and Tabata Development Fund. Community participation and self-help is common to all.[29]

Self-help activities undertaken included community mobilisation by *Mtaa* and Ten Cell leaders and/or supply of physical labour, contribution in cash and materials such as land for infrastructure installation and building materials and cleaning of open wells. Construction of wells by households as well as those owned by vendors was done on a self-help basis. Self-help constructions appeared to lower the overall project costs because work was organised and supervised by the *Mtaa* leaders or family members with no pay, land for installing infrastructure and materials was also obtained free of charge.

Self-help Contribution:

• Community management and organisational activities offered by the *Mtaa* leadership. Also ten cell leaders voluntarily managed to supervise water development initiatives e.g. by identifying space for locating public wells.
• Households contributed funds[30] and labour in construction, operation and management of water wells and water tanks.
• Local leaders' supervision of operation and maintenance of the community water supply.

- Provision of land for digging wells and installing water tanks by the landowners.

Limits to Self-help

Inadequate capacity of the *Mtaa* leadership to co-ordinate community activities was found to be one of the obstacles in enhancing self-help initiatives in the two studied areas. The respondents associated the capacity of the *Mtaa* leadership with lack of animation and poor communication with the community that seem to have undermined self-help activities in the two settlements.

About 40% (12 out of 31) of the respondents in Tungi alleged that they had not been mobilised for self-help activities. It is however highly unimaginable that so many residents in Tungi were unaware of the need to contribute towards improvement of the water supply. It is probably not wrong to speculate that the differing positions held by the respondents depicted varying/differences in the community towards solutions to the infrastructure deficiency in Tungi. These may also show discontent and heterogeneity in the community. The foregoing is one of the challenges which underline initiative/attempt to improve basic infrastructure in settlements, which exhibit widespread differences.

Informal discussions with various residents in the two areas, showed that the residents were informed and were aware of self-help activities but could not contribute voluntary labour, because they were employed or working on their farms. Inability to contribute could also be attributed to poverty among the households. This ought to be a barrier to participation in self-help activities because most times poor households are engaged in economic activities which last long hours with low pay per hour or per day. The income they earn might not be enough for the household expenditure and to contribute to self-help activities. Also, they find little time to participate in self-help activities.

Normally, local communities, no matter how poor they are, have some capacities for self-help improvement.[31] Therefore, communities should, in principle, get assistance only after they have exhausted their own initiatives. External assistance ought to be a form of help towards self-help, which would be intended to provide initial stimulus.[32] As noted in the two settlements, external funds were used to construct wells that were later handed over to the community for operation and maintenance. However, the granting of the funds by external agents without exhausting the community's own contribution and commitment for the construction of wells seems to have negated local initiatives. This is in the sense that the community became more dependent on external support. External support seemed to have reduced the impetus of the residents on self-help activities. From the interviews it was noted that because some residents believe that the wells belong to the government, people should be paid to guard them.[33] In general, external assistance and linkages were essential to support the community initiatives.

The Necessity of Horizontal and Vertical Linkages

Horizontal and vertical links have developed in the course of improving water supply in the two settlements. It is the linkages, which seem to have facilitated exploitation of human, financial, material, organisational potential within and without the settlements.

Horizontal Links

Five main actors have forged links with the *Mtaa* leadership within the settlement. These include the residents, professionals and *fundis,* water vendors, political parties, and formal and informal NGOs. These actors were and are involved in various activities in the water provision process. The main activities include leadership and organisation, mobilisation of resources, construction of wells especially traditional ones, water supply and selling and operation and maintenance of wells.

As regards the horizontal linkages there are strengths and weaknesses in the provision process. Strengths include an existing organisation or institution that could be built through the reconciliation of formal and informal norms and regulations, trust in the leadership, co-ordination and management of the water supply. Others include existing local resources and voluntary contributions such as labour and materials, private investment and collective action. Threats in the provision process were noted to include poverty among households, lack of transparency in community activities and hidden interests such as for political gains by both the *Mtaa* leadership and the political parties.

It has also been observed that there are various horizontal links that have potentials for infrastructure provision to the fringe settlements. The links have facilitated *Mtaa* (institution) building and development, community organisation, mobilisation and utilisation of local resources including knowledge and skills, access to safe water, etc. Lack of trust, rules and regulations and abiding to commitments may create disharmony in the links established, thus threatening sustainability in the water provision process.

Vertical Links

A local infrastructure initiative that is being implemented in an environment of increasing poverty often would require external support especially technical assistance in order to succeed in the delivery of the service[34]. Partnerships or collaboration among the public (state), private sector, the community and donors are essential. There are several links that have been forged between the community and upper departments of the central and local government and/or other institutions outside the settlement. The *Mtaa* leadership and individuals including water vendors forged links in technical, financial and administrative support. These were important because it was a way of facilitating low-income groups' right to act, organise and make demands. Making people aware of their own capacities and resources can help increase the options available to them.

Links between Mtaa Leadership and Ward Level Office

The Sub-ward has strong links with the Ward office especially in information dissemination to and from the residents to higher organs of the government.

All directives from the Municipal or City council for example keeping the environment clean, boiling water before drinking, implementing self-reliance activities and collection of development levy go through the Ward Executive Officer to the Sub-ward leadership. Because some of the DCC and central government administrative and management units such as community development, education co-ordination, health services, agricultural extension services, and Court services are available only up to Ward level, the need for a link between the two systems is very critical. Officers in these units provide supportive services to the Sub-ward level. However, the needed support was not frequently being provided due to inadequate "capacity" at the Ward level particularly in terms of logistics and finance. Subsequently, the link between the two exists but this cannot be said to have been very functional. Often due to lack of incentives such as transport and attractive remuneration to staff, they do not visit projects in the *Mtaa* areas.

Links between Mtaa Leadership, the Community and Donors (e.g. Lions Club and the UNDP)

The Lions Club, an NGO with a branch in Dar es Salaam granted funds to support Tungi community in the construction of five hand pump wells. A resident in the area who is also the chairman of the Water Management Committee initiated the link between the NGO and the community. Through the implementation of this initiative a link was also established between the Community leadership and the Water Resources Institute who constructed the wells. The Local Initiative Facility for the Environment (LIFE) of the UNDP on the other hand also donated US $12,300 to Yombo Dovya community to support them in the construction of five wells. As discussed earlier, the Yombo Dovya initiative was not sustained because water was fetched free of charge and thus no effective operation and maintenance was put in place. The initiator of this project was the LIFE facility.

Links between the Water Management Committee, DAWASA and the Water Resources Institute

DAWASA with its technical competence in water provision was contracted by DCC who received the funds from UNDP to construct the water wells in Yombo Dovya that belonged to the Sub-ward. They have since constantly supplied the two settlements with chemicals for water treatment. In addition to this, they also support the vendors with drilling of wells and repair of pumps albeit informally. DAWASA has also been training the water vendors and the Water Management Committee members in water treatment. The Water Resource Institute's Instructor living in Tungi facilitated the link between the Water Resource Institute and the community. Students and staff constructed the five wells. The three national organisations (DCC, DAWASA and WRI) acted as focal institutions that are vital in supporting the water infrastructure provision at the local level.

The Stakes: Poverty Reduction

The water initiatives in the two settlements had the objective of improving water supply as a basic need for residents in fringe areas. The provision of the service led to improved living conditions of the people (most of them are poor) especially because the case has shown that employment and household income had been acquired in the process of providing the water. Water was sold to generate income. Collective action and participation in the provision process had increased the options available to low-income groups and thus ensuring constant supply and hence keeping down the price of water per litre. Above all, competition that had evolved in the process has ensured the supply of water to meet different demands based on affordability levels of the different income groups in the two areas.

While quality of service depends on the source, the suppliers especially the vendors have been trying to adopt appropriate technology to improve upon the quality of the service provided to meet demand. It is thus shown that sustainability may be achieved in the provision of water supply to fringe areas through participatory and cost recovery measures, but it is threatened by lack of operation and maintenance.

Access to Potable Water

Households get access to water at affordable prices because of adequate supply. Most (70%) of the respondents access water supplies within 200 metres distance. Consumers spend an average of 15 minutes to fetch water. On average, only 3 minutes are spent at the source. Comparing these figures with those in Buguruni or Tabata, the figures indicate that the water supply in the settlements is very satisfactory although it needs improvement. The average time of fetching water in Buguruni is 30 minutes if water is available in the taps, about half of the residents use one hour or more to fetch water. The average time taken to fetch water in Tabata is 18 minutes. Thus, the situation in these fringe settlements is better than that in the two areas of Buguruni and Tabata.

Willingness to Share Costs

About 90% of the respondents expressed the willingness to contribute towards the capital cost for the improvement of water supply in their settlements. The rest (10%) did not indicate any contribution but it has to be understood that water is still available free of charge to some households in the fringe areas. The latter therefore see little or no point in contributing towards improvement of the system[35]. Other settlers who could not contribute cash were willing to contribute physical labour in addition to voluntary leadership.

Ensuring Quality of Water

The quality of water supplied depends on the source. Water from shallow wells is salty. The quality of water in the two fringe settlements has been improved in two ways: by providing chemicals to treat the water from shallow wells or by digging deep wells. Chemicals are obtained from the government and distributed by the

Water Management Committee. Private operators on the other hand ensure a better quality of the water supplied by using the services of technicians from the government.

Poverty Reduction

Three of the major causes of poverty include the limited access in the use of local resources, be it land, labour, finance etc.; limited access to basic infrastructure such as safe water, access roads, health and education; and lack of community organisation which hinders participation and decision-making of the poor. Poverty arises because the poor cannot efficiently use the resources around them to generate adequate income. As noted in the discussion above, there are several indicators that have emerged that can ascertain poverty reduction. The main ones are enabling community participation in water supply and management, availability and improvement of quality of water for better access and use of water as an asset for employment and household income generation.

Lessons Learned

It is clear that there are various potential resources available at the community level in fringe areas and if these are mobilised, they can be used in providing basic infrastructure like water. Other local actors however, must be deployed to supplement the efforts of the community. Mtaa gets knowledge, skills, materials and finance through creation of collaboration networks. In fact, acquiring knowledge and skills has been through learning-by-doing where in some situations mistakes have been corrected.

The role played by the public was in giving advice and technical support; the private sector in investments and operation; the community in organisation and co-ordination and the donors in financial support. These were inevitable in meeting the basic needs of the growing homogeneous populations. Although the main objective of the external and internal actors was to enable the communities in both areas to have access to potable water, it appears that leaders in the communities had varying interests, both political and economic. At the beginning, in the early 1990s, although the government policy was to ensure community services such as water supply was made available through self-reliance, the leaders were trying to safeguard the interest of the party that was free water for all without knowing its implication. In this regard, the purported political gain superseded sustainability of the provided basic service leading to collapse of a system.

Making people aware of their own capacities and resources can help increase the options available to them. Professional advice can increase the choices further, but successful professional intervention requires that the value of such intervention is recognised and accepted by the low-income households.

It has been noted that the creation of a new institution e.g. the Water Management Committee to participate and ensure safe water was available in the settlements was met with different experiences. While the organisation rules and procedures followed allowed collaboration with the private sector to ensure water was available to the residents, it was taking long to adapt to institutional change especially in providing water at a charge. The inability of the Water Management Committee to provide sufficient water gave rise to water vendors. Although there was improvement in the water supply in both quantity and quality in both settlements due to the competition which had evolved, there will be future control of the water resources in the areas. It is indisputable that participation of the different actors brought in invaluable resources, however, the systems in place in both areas need to be better co-ordinated and more importantly, reviewed to ensure sustainability in the water provision process.

Without cost recovery a project collapses. The pre-condition for a service to be sustainable and efficient is that there must be an effective cost recovery mechanism. In the present case it has been clearly observed that even with financial assistance to build the infrastructure, the project will collapse if there is no mechanism for recovering costs for operation and maintenance.

Again, costs can be kept down by organising and managing the construction, operation and maintenance by the people themselves. This can be done at household level or by having elected or appointed members of the community to take on tasks for which they are not paid. There is however recognition of the limitations of the elected or appointed representatives from the poor households, since they have inadequate incomes and limited time to devote to organisation and management. The costs were kept down by having free or informal technical support (through negotiating special deals) from formal institutions like DCC, DAWASA and the Water Resources Training Institute. These formed the focal institutions required to support local initiatives.

It is important to note that in order to adopt higher technology, household saving is not enough thus, it must be supplemented with external resources support e.g. from friends, NGOs, the government or international organisations. The vertical and horizontal links created were inevitable to facilitate local and external inputs into the provision and improvement of water supply in the settlements. The collaborations ensured that the community could use locally available resources to supply safe water. However, the donations of capital funds from external sources to build the wells seemed to be a one-time input that did not involve a continuous follow-up from the collaborators to evaluate progress or constraints in order to change rules and procedures in the water provision so as to ensure sustainability.

It might be that they (donors) did not wish to be involved in the politics of the water provision that included free water service to the people. It was believed that the residents were poor and thus were not able to pay for the water; this was a misleading policy that the donors did not like to be involved in. In fact, the consequence was the collapse of the initiative that led to the people resorting to vendors and/or traditional wells. This, no doubt, led to higher expenditure on water and increase in diseases because of use of contaminated water from shallow wells.

Private investment in water provision by vendors has proved to be effective and sustainable due to the element of cost recovery that was built in. The sub-contract of water management in Tungi has also brought in the same sustainable results due to its innovative idea of selling water for the purpose of getting revenue for operation and maintenance.

External and internal collaboration is necessary to have improved water supply in the settlements. However, it was noted that vertical and horizontal links evolve through various interests, some are hidden some are open. Linkages should not be one-time help but a long-term assistance to ensure that processes that have taken off do not collapse. It was also noted that without cost recovery for operation and maintenance of infrastructure such as water supply, the service could not be sustained. Allowing competition in infrastructure provision ought to assure high quality service while the cost recovery imbedded in the provision allows a step-by-step improvement of the system according to demand. The Mtaa leadership in collaboration with informal institutions could effectively co-ordinate various actors who have interest in improving infrastructure such as water supply in an informal settlement. Often without self-help or voluntary leadership, various functions may not be performed.

Conclusion

With safe and clean water in a settlement, the living environment is improved. Water as a commodity is a source of revenue for operation and maintenance; as technical infrastructure it generates employment and income. Thus, water is both a social service and an asset that could be transacted to reap household income and revenue not only for operation and maintenance of provided basic infrastructure, but also for facilitating a step-by-step improvement. Private operators seem to be more efficient in managing wells and pumps. The private sector seemed to have strong informal links with public institutions such as the DAWASA technicians. The latter provide services against some agreed payment, which was informally paid. On the other hand, the private sector abides to the principle of cost recovery for operation and maintenance.

For the Mtaa to be able to discharge its duties efficiently, they needed to have more finance (economic empowerment) through the decentralisation of infrastructure provision, development control and property tax collection. Property tax collection is a centralised city activity and the city has no capacity to reach out to fringe areas. If collected by the Mtaa leadership, it could be used to create a revolving fund or provide basic infrastructure to the community. There seems to be a need to decentralise property tax collection and provision and management functions of technical infrastructure to local levels in order to improve demand-responsiveness. However, varying interests, if not checked, may direct funds to other areas as experienced in Tungi and Yombo Dovya.

As house and population densities increase with continued use of pit latrines, the ground water is likely to be contaminated. The need to closely link/co-ordinate basic infrastructure improvement programmes with population increase is essential. Low-cost water supply options are feasible up to a certain density level only; beyond this, other options have to be explored. For a more efficient, effective and sustainable water supply in the fringe settlements, a review and improvement on the existing models is needed. It must also be noted that simply because the settlements are not on the water transmission line, the major source of water supply in the area in the short and medium term would continue to be ground water that need to be monitored closely for sustainable exploitation.

In order to control/check density (housing) for the purpose of operating an affordable and sustainable system in fringe areas, there is the possibility of using the Mtaa leadership and its sub-committees which can make negotiations with plot/house owners for land to install community services. The case in Dar es Salaam has shown that the Mtaa leadership is accepted and trusted by the people and has capacity to co-ordinate infrastructure improvement in fringe areas if given adequate assistance through partnership arrangements.

Notes

1. Kyessi, A.G. (1990, MSc Thesis). "Urbanisation of Fringe Villages and Growth of Squatters: The case of Dar es Salaam, Tanzania". ITC, Enschede & Kombe, W.J. 1997b. "Regularising Urban Land Development during the Transition to Market-led Land Supply in Tanzania", in Kombe, W.J. & V. Kreibich (eds.). Urban Land Management and the Transition to a Market Economy in Tanzania. Dortmund, *SPRING Research Series no. 19*, pp. 29-48. Analysing urban land development during the transition to market-led land supply in Tanzania, Kombe asserts that the habitate strategy predominates in the many informal settlements. It has at least helped in holding the land price low and thus making housing land affordable to relatively more people than would be the case with serviced land. Service-as-you-inhabit approach appears flexible and largely blends with the desire of most would-be-home seekers to resolve the most daunting problem in the course of acquiring a house first, namely the housing land acquisition puzzle.

2. Kyessi, A.G. 2001. "Community-based Urban Water Management under Scarcity in Dar es Salaam, Tanzania", in *The Journal of Building and Land Development, Vol. 8*, no. 1-3, pp. 28-41.

3. As with the case of Yombo Dovya, elders in the area explained that the area started to develop fast starting from 1973 when the food production campaign to curb famine was launched. Many people who were searching for farming land settled in the area because of the availability of fertile land.

4. Discussion with elderly people of Yombo Dovya held on 27 May, 1999.

5. Kyessi, A.G. (1990, MSc Thesis). *"Urbanisation of Fringe Villages and Growth of Squatters: The Case of Dar es Salaam, Tanzania"*. Enschede: ITC.

6. Kombe, W.J. 1995. "Formal and Informal Land Management in Tanzania. The case of Dar es Salaam City". Dortmund, *SPRING Research Series, no. 13*.

7. Mwamfupe, D.G.(1994, PhD Thesis). *"Changes in Agricultural Land-Use in the Peri-Urban Zone of Dar es Salaam"* University of Glasgow.

8. Kyessi, A.G. 2002. *"Community Participation in Urban Infrastructure Provision: Servicing Informal Settlements in Dar es Salaam". SPRING Research Series, No. 33*, p. 4001. Dortmund: University of Dortmund.

9. Kyessi, A.G. 2001. "Community-based Urban Water Management under Scarcity in Dar es Salaam, Tanzania", in *The Journal of Building and Land Development, Vol. 8*, no. 1-3, pp. 28-41.

10. The most commonly used way to measure poverty is based on incomes or consumption levels. The minimum level is usually the poverty line. Per capita expenditure of less than US $ 0.75 a day is categorised as a population living in absolute poverty.

11. Even though these figures show that the residents in studied areas are within the absolute poverty group, their living pattern does not seem to support this, many can afford basic needs including food, shelter and basic medical attention. In this case, there seems to be some sources of income which could not be established.

12. Kombe, W.J. 1995. *"Formal and Informal Land Management in Tanzania: The case of Dar es Salaam City"*. Dortmund, *SPRING Research Series, no. 13.*

13. These are land use plans prepared and approved by the Ministry of Lands and placed at the Dar es Salaam City Council for implementation.

14. Kyessi, A.G. 2001. "Community-based Urban Water Management under Scarcity in Dar es Salaam, Tanzania", in *The Journal of Building and Land Development, Vol. 8,* no. 1-3, pp. 28-41.

15. *Mtaa* or urban sub-ward is the smallest unit of administration in local government functions in urban areas in Tanzania.

16. A clause in the Local Government policy existing during the time of fieldwork allowed the Chairman to appoint only one resident as a Secretary of the *Mtaa*. After being amended, the current clause allows the residents to elect its Chairman, the Secretary and five advisors, of which, two of the advisors must be women. This new procedure was adopted in the sub-ward and village elections in September 1999.

17. Discussion with one of the members of the Water Management Committee, Mr Kazimoto Gaudence, January 1, 2000.

18. Discussions with owners of private water tanks on January 1, 1999.

19. The space for one well is about two square metres and it has to be given free of charge as a household contribution towards construction of the well. CCM Ten Cell leaders co-ordinated the negotiations during the time of single party democracy.

20. The actual amount contributed and collected by the Ten Cell leaders could not be communicated to the author indicating perhaps that there was no control of community funds.

21. During the period of single party democracy in Tanzania, the practice was that the ruling party, CCM, looked after all community projects. Their leaders had two roles - political and government leadership. There was no separation of these powers. At the local level it was believed that this system would lessen obstacles that might have existed in the smooth management of community development. In constructing the wells, technical assistance was sought from DAWASA.

22. The average household size is 5.2 persons in both Yombo Dovya and Tungi.

23. Kyessi, A.G. 2001. "Community-based Urban Water Management under Scarcity in Dar es Salaam, Tanzania", in *The Journal of Building and Land Development, Vol. 8,* no. 1-3, pp. 28-41.

24. Anzorena, J., Bolnick, J., Boonyabacha, S., Cabannes, Y., Hardoy, A., Hasan, A., Levy, C., Mitlin, D., Murphy, D., Patel, M.S.; Satterthwaite, D. & Stein A. 1998. "Reducing Urban Poverty: Some Lessons from Experience", in *Environment and Urbanization, Vol. 10,* no.1, pp.167-186.

25. Anzorena *et al.* 1998. *Reducing Urban Poverty: Some Lessons from Experience*, write that in a single project, neighbourhoods within the same project area choose different ways of organising. For instance, in Orangi in Karachi, each lane within this huge informal township chose its own way of organising and generating cost recovery when developing sewers. The same could be related to the water producers in the two settlements who are meeting the demand of the different income groups of the residents.

26. Informal discussion held on February 5, 2000 with Mr E. Mundo, Chairman of the Water Management Committee who was the coordinator of the construction of the five wells in Tungi. Mundo explained that once the wells were constructed, by students from the Water Resources

Institute in Dar es Salaam City in a scheduled field practical, they were handed over to CCM. Water was provided free of charge in the era's ideological spirit of providing free basic services to everyone.

27. Discussion held on May 3, 1999 with the *Mtaa* Secretary. It was noted that the decision was made by the Water Management Committee in 1996.

28. Deutsche Gesellschaft für Technische Zusammenarbeit (GTZ) 1997. *"The World of Words at GTZ".* Eschborn.

29. Kyessi, A.G. 2002. *"Community Participation in Urban Infrastructure Provision: Servicing Informal Settlements in Dar es Salaam". SPRING Research Series, No. 33,* p. 4001. Dortmund: University of Dortmund.

30. The exact amount of funds contributed could not be obtained from the leaders and this is linked with absence of a ledger for recording voluntary contributions. It is in fact one of the complaints raised by the respondents, that of lack of transparency in conducting community activities.

31. Schübeler, P. 1996. *"Participation and Partnership in Urban Infrastructure Management".* Urban Management Programme, The World Bank, Washington, D.C.

32. Deutsche Gesellschaft für Technische Zusammenarbeit (GTZ) 1997. *"The World of Words at GTZ".* Eschborn.

33. Discussions with three members of the Water Management Committee held on February 1, 2000 in Yombo Dovya. The *Mtaa* Chairman and some of the residents interviewed supported the views of Water Management Committee.

34. Kunfaa, E. 1996. "Sustainable Rural Health Services through Community-based Organisations: Women Groups in Ghana". *SPRING Research Series, no. 19.* Dortmund. Hasan, A. 1997. *"Working with Government".* The story of OPP's collaboration with state agencies in replicating its Low Cost Sanitation Programme. Karachi:City Press.

35. The maximum contribution that was cited for improving water supply was TShs.10,000 and 5,000 in Tungi and Yombo Dovya respectively. The minimum was TShs. 200 for both areas.

Water Vending in Dar es Salaam, Tanzania

Marianne Kjellén

Introduction

The general model of urban water supply pictures water utility delivering water to all the residents of a city. However, in developing countries, it is most often the case that only a part of the population actually receives this service. In practice, access to water distribution networks is restricted. This paper explores how water vending caters for onward distribution to many of those who are not connected to the piped system. It reviews issues of pricing and the potential for more structured collaboration between vendors and the 'official' water system.

Dar es Salaam is the major city in Tanzania. It has a population of some 2.5 million[1] and it is divided into three municipalities; Kinondoni, Ilala and Temeke. Poor urban management, rapid urban growth and low levels of investment combine to produce a problematic situation with regard to infrastructure, public services, urban health and environment.[2-4] The 2002 Census recorded 92% of the households in Dar es Salaam as having access to water; 76% in the form of piped water supplies and 16% by way of other "protected" sources of water. The three municipalities differ greatly, with piped water estimated to reach some 94% of the residents in Kinondoni, 72% in Temeke, and only 49% in Ilala.[5]

The higher coverage in the northern-most municipality of Kinondoni can be partly explained by its favourable location with regard to piped water. The transmission pipelines leading water to the city from the main water works on the Ruvu River runs through this relatively wealthy area, where also a larger share of the population can afford to connect. Piped water connections are expensive in Dar es Salaam, due to the city's underdeveloped distribution system, which forces private connection lines to be extremely long.[6, 7] Moreover, Kinondoni, encompassing most of the city's previous colonial residences, is generally favoured with regard to the provision of infrastructure. In Ilala and Temeke, the lower coverage reflects both private poverty, i.e. inability to connect, and lack of public investment, i.e. limited extension of distribution system. Both these municipalities encompass most of Dar es Salaam's working classes, with the originally planned areas for 'Africans' and the largest concentration of industries in the port, railway and airport areas. The low coverage of piped water services to households in these areas is somewhat compensated for by a better position with regard to groundwater. These southern and western parts of Dar es Salaam drain the nearby Pugu Hills.

It should be noted that throughout Dar es Salaam, households usually rely indirectly on piped water. The 2000/01 Household Budget Survey found that while only a third of the households had private connections, another 46% also relied primarily on piped water that they acquired from their neighbours.[8] The *Drawers of Water II* study indicated that water vending has increased over time. In some parts of Dar es Salaam, 60% of the households use vendors as their primary source of water. As secondary source, when piped households' regular supply channels do not function, water vending is even more important.[9]

Tanzanian water sector has been greatly reformed since the 1990s. During that decade, water and sewerage services in urban areas were combined and greater independence from the Ministry of Water was instituted. This reorganisation also laid the basis for subsequent privatisation and commercialised operations. In Dar es Salaam, the National Urban Water Authority, which had a national mandate but only operated in Dar es Salaam and neighbouring areas, was merged with the Dar es Salaam City's Sewerage and Drainage Department to form the Dar es Salaam Water and Sewerage Authority (DAWASA) in 1997.[10] The DAWASA privatisation was launched immediately in 1997, but following a lengthy process of bidding, only in 2003 could a private operator take over.[11]

City Water Services was a joint venture between the British-German consortium Biwater-Gauff Tanzania Ltd (BGT) and the Tanzanian company Superdoll Trailer Manufacture (STM). With great expectations City Water started operations 1 August, 2003, but less than two years into the ten-year lease, the Government of Tanzania cancelled the deal.[12] The government was not satisfied with City Water's performance, and City Water was dissatisfied with the operating conditions. The problems described in the proceedings of the international arbitration trial were: 1) a poorly prepared bid by BGT, 2) not making use of the 'enhanced monitoring period' for re-calibrating base figures, 3) numerous management and implementation difficulties, resulting in City Water not generating the income which had been foreseen, 4) coming to the situation where City Water could not continue without a fundamental renegotiation of the contract, and finally, 5) when renegotiation took place, this failed.[13]

A new, government-owned, water company was quickly formed, and the Dar es Salaam Water and Sewerage Corporation (DAWASCO) started operating on 1 June, 2005. DAWASCO still operates Dar es Salaam's water system, and also implements the World Bank financed rehabilitation project that was designed to accompany the privatisation deal. A smaller licensing authority, retaining the name DAWASA, and a regulatory authority, the Energy and Water Utilities Regulatory Authority (EWURA), are in charge, respectively, of contracting and monitoring of the operations.

This exploration of water businesses in Dar es Salaam forms an integral part of a PhD project on institutions for water provisioning.[6] Field work was carried out in one to two-month sessions in 1998, 1999, 2000 and 2004, in areas where the municipal water supply is problematic. A questionnaire survey among water vendors was carried out during 1998 and 1999 sessions, the results of which have been published both in English[14] and Swahili.[15] Further informal visits to vendors and vending locations to check on prices and conditions were done during subsequent sessions and in 2006. Other parts of this research project explored households' situation in different parts of the city and the local and international forces involved in the privatisation process.

Water Vendors – the Retailers

This section reviews private (commercial) water re-distribution in Dar es Salaam. It concentrates on distributing vendors, using manual labour or tanker trucks for the delivery of jerrycans or tankloads to consumers' homes, and households re-selling water obtained from their own connections or boreholes to consumer households collecting for themselves, as well as to distributing vendors.

Households (re-)selling Water

As mentioned, most bulk water is supplied to the city by means of the utility-operated pipelines. But since few households are directly connected to the distribution network, many of those that are connected run informal water kiosks where they sell water to their neighbours. Connected households generally pay monthly bills, with a flat rate based on the estimated water pressure in the geographic area. Increasing numbers of households, however, are being metered, in accordance with the metering policy introduced in 1997,[16] but only gradually implemented.

Usually, water is delivered through a rubber hose extended into the street, or customers may tap water inside the compound. Where pipe-water is rationed, those aiming to sell on a regular basis must construct storage facilities in order to be able to sell water during 'off-turns'. Also, in order to make sure that the storage facilities are filled; it may be necessary to connect a booster pump in order to suck (the low pressure) water out of the pipe system. Many households, particularly in peripheral areas, also sink wells in order to match sources and ensure a continuous supply. As long as there is electricity, they are able to continue pumping. However, not all areas have groundwater. Where there is, it is often saline, and given the generalised use of pit latrines in Dar es Salaam, it is likely to contain faecal contamination. One household that was engaged in the re-selling of water by the bucket estimated its sales to Tanzanian shillings (T.Sh. 20-25,000 roughly US$ 20-25. *An exchange rate of 1,000 Tanzanian shillings per US dollar is used throughout this paper, representing a rough average for the going rates in the early 2000s*) per day in the dry season, and to about T.Sh. 5,000 (US$ 5) per day in the wet season.[17] The family had constructed a large tank, and when piped water was insufficient, it was topped up by pumping from their own well. The monthly water bill paid to DAWASA amounted to T.Sh. 15,000 (US$ 15). Although significant earnings are needed to cover the investments in tanks, wells and pumps, water selling is clearly a profitable business.

In most areas of Dar es Salaam, the price for resold water is T.Sh. 20 per bucket or jerrycan (of approximately 20 litres), equivalent to US$ 1 per cubic metre. The (relatively few) DAWASA/DAWASCO-operated water kiosks also sell at this price. However, as part of the ongoing rehabilitation project, some 250 water kiosks[18] are to be constructed in Dar es Salaam, and licensed to individual households.

Water Distribution by Pushcarts

Water vendors using hand-driven carts operate mostly in the less well-off western and southern parts of the city, where water is rationed or where piped distribution systems are deficient or non-existent. These small-scale water vendors, like most households, often buy their water from water-selling households or water kiosks, and then deliver the water by the container to consuming households and small-scale businesses.

Pushcart vendors tend to be younger men with some schooling, seeking ways to earn an income. Most carts carry six or seven plastic jerrycans. The major problem that vendors face, especially when getting started, is the toughness of the job. During the survey, many complained of pains in the chest and joints, and that they often fall sick with fever. The earnings are generally low, and vendors claim they at times go hungry. Calculations found average earnings to arrive at less than a minimum wage.[14, 15]

Commonly, 20 litres are delivered for 100 shillings (equivalent to US$ 6.25 per cubic metre), but the price varies depending on the accessibility of the area and the distance from the water source. This price is considerably higher than what connected households pay to the water company or what is paid at water kiosks. The mark-up by vendors is motivated by the considerable work implied by the physical delivery of the water. It should be noted that vendors do not have any special treatment from the water company, and hence do not have access to any cheaper water than consuming households. Only, vendors may at times get seven jerrycans of water for the price of six. Thus, included in the vendors' prices are the DAWASA/DAWASCO flat rate, the mark-up by the re-selling household, 'capital costs' for pushcarts and jerrycans, as well as the vendor's own compensation for the effort.

The process of price formation varies with different levels of scarcity conditions. During regular operations, or 'normal' water scarcity, vendors appear to be in consensus about what is the going price in each particular area. The prices are historically given, and continuously reproduced by the vendors themselves as they conform to market conditions. Hence, during normal times, the vendor market functions in the vein of 'old institutionalists' who see prices as providing norms or conventions.[19, 20] That is, vendors are price takers, like agents operating under a structure where each is unable to affect the price or perceive their own part in the whole.

By contrast, during 'extreme' scarcity, vendors become 'opportunistic self-seeking atomistic agents' that try to get as high prices as possible. Prices are then produced in the meeting point of supply and demand, and the water vending market operates in a way which is consistent with neo-classical views and 'new institutionalists'.[21] During such circumstances, a 20-liter jerrycan of water can be sold at T.Sh. 500-700 shillings per container,[14] equivalent to US$ 25-35 per cubic metre. Vendors also move out of their normal selling areas, in order not to spoil existing customer relations, and gravitate towards the centre of town where prices are said to be higher (moreover, during crisis conditions, water consumers also venture out of their dwelling areas to hustle for water).

The initial investment required to enter into pushcart water vending is very small. The pushcarts, and even the jerrycans, can be rented on a daily basis. Hence, there are virtually no barriers to entry, and the number of vendors that operate varies with market conditions.

Vendors are often accused of delivering water of dubious quality. Indeed, a vendor wanting to 'steal' customers from his competitors may help spread such rumours. During the survey, however, vendors appeared to take significant precautions to safeguard water quality. Jerrycans were claimed to be regularly washed, an important action for water quality concerns, but probably even more crucial for successful water sales. Vendors were also conversant with the quality of water at different sources and how it varied seasonally. Indeed, the commercial pressure upon vendors appears to function as a check on water quality, as "vendors interviewed seemed more worried about the scarcity of clients than the scarcity of water."[14] That vendor-delivered water in Dar es Salaam is of reasonable quality, or at least not of systematically inferior quality, is confirmed by the findings in the *Drawers of Water II* study, where households relying on such water had lower than average diarrhoea prevalence.[9]

Water Distribution by Tanker Trucks

In low-density areas with well-off residents but poor water infrastructure, trucks play an important role in water distribution. Tanker delivery has increased from virtually nil in the mid-1990s to a major business a decade later. By the late 1990s, there were probably not more than twenty tankers operating on the market, which, however, was to explode during subsequent years, and both freshwater and sewage suction tankers are highly visible on the streets of Dar es Salaam. [22-24] Along with the enormous increase in the number of boreholes drilled by the government as well as by private persons during the late 1990s,[25, 26] tankers that sell groundwater came to operate throughout Dar es Salaam.

The major tank-water vending point is located by the hydrant at *Sayansi* (the Tanzania Commission for Science and Technology offices) in Kinondoni District. This is not far from the major area of tanker sales along the Indian Ocean coastline, where most of Dar es Salaam's wealthier inhabitants live. While the supply could be expected to be ample and reliable, as it is not far from the Lower Ruvu Transmission Line, it is actually very deficient in some areas. Service lines are of PVC and leak profusely, resulting in very low pressure.

The water prices paid by tankers to the water company used to be very high, with a truck-load of water costing beyond T.Sh. 50,000 (equivalent to over US$ 5 per cubic meter) during the late 1990s. Following complaints from tanker operators, and the fact that many were filling up their trucks elsewhere, DAWASA lowered their charges. By 2004, the considerably lower tariff, T.Sh. 10,000 per tank (US$ 1 per cubic meter), was applied also by City Water Services. The lowering of input prices resulted in a further expansion of tanker sales, with more tankers on the market and lower end-user prices; some T.Sh. 30-35,000 (US$ 30-35) per tank compared to almost the double in earlier times.[6] The reduction of sales prices in response to reduced input costs as well as the increasing number of tankers in operation, suggest effective competition on this market.

Water Prices in Dar es Salaam

Different types of water distribution modes vary in their price levels and forms of payment. Table 1 displays the prices of different sources in Dar es Salaam in the early 2000s. As mentioned, connected households most often pay flat monthly bills, implying that the actual volumetric price fluctuates according to the actual volumes consumed. The utility tariff underlying the monthly charge (at the time of privatisation in 2003), however, amounts to less than half of the litre price paid by households collecting by the bucket at resellers or kiosks. Water delivered to the home by distributing vendors cost several times more, as it also covers a premium for beyond-pipe water transport. That is, the incremental effort required to deliver water to a given consumer depends on their distance from the source and the presence of other consumers en route. Also, road conditions and elevation affect prices charged by vendors.[6, 14]

Approximate Volumetric Prices of Water from Different Supply Modes in Dar es Salaam, early 2000s					
Source / Mode	Characteristics /requirements	Form of Payment	Nominal Price	T.Sh. per litre	US $ per m3
Piped water – own connection	Requires connection to premises	Monthly lump-sum, generally flat rate	T.Sh. 451 per cubic metre (2003)	0.45	0.45
Private (household) water re-seller	Buyer carries from point of collection	Cash by container, or neighbourly agreement	T.Sh. 20 per 20-litre bucket or jerrycan	1.00	1.00
Tanker truck	Requires large storage reservoirs on users' premises	Invoiced or cash by tank-load. Often negotiable.	T.Sh. 30 - 45,000 per 10,000-litre tank	3.00 - 4.50	3.00 - 4.5
Pushcart water vendor	Delivered to house (into storage vessel)	Cash, by container or other agreement. Prices vary with distance from source.	T.Sh. 70 - 200 per 20-litre jerrycan	3.50- 10.00	3.5 - 10

Note: US Dollar conversions made at the rate of T.Sh. 1,000 per US_$ 1:00.

Data source: Kjellén, Marianne, (2006, PhD dissertation) "From Public Pipes to Private Hands: Water Access and Distribution in Dar es Salaam, Tanzania," Department of Human Geography, Stockholm University.

Potential for Structured Collaboration between Vendors and Utility

To remedy the current inequities and problematic water situation in Dar es Salaam, a more efficient and accessible pipe-water distribution system is required. But, even with the ongoing rehabilitation project, Dar es Salaam is a long way from universal piped water services. Distributing vendors thus need to be recognised as a complementary means for water distribution. Indeed, they constitute a mobile extension of the piped network. With a better-delineated cooperation between water vendors and the network operator, such an 'extension' could be made more efficient in meeting water needs of Dar es Salaam residents.

Water vending differs from piped supply in that it is much more flexible. Piped systems are among the most expensive type of societal infrastructure and have a pay-back time of up to a hundred years. Both truck-delivered and, in particular, pushcart-delivered water requires much less investment. This flexibility allows vendor systems to be 'installed' virtually over-night, whereas piped systems have a very long gestation time. In addition, vendor systems are open to competition in a way that piped systems are not. This provides an informal check on the price levels and water quality.

Notwithstanding, vendor-delivered water is notoriously expensive. The price that the consumer has to pay is much higher than what is paid for piped water directly from the utility, see table 1 above. There are several reasons for this:

• The vendor needs a margin to make the delivery worthwhile. As mentioned, water conveyance by feet or wheels is cumbersome and hence very expensive. It appears that the present margins for pushcart vendors are at the level equivalent to a minimum wage, or less. Vendors operate in areas where water deliveries are more difficult. That is, where the sources are far away, water conveyance becomes more expensive (even though utility tariffs rarely differentiate prices for delivery difficulties, a high-lying area is more expensive to service than in a low-lying one, also for piped water deliveries).

• Water vendors have no preferential treatment as could be expected in a wholesaler-retailer system. Vendors purchase water on the 'open market' at the same prices as consumers (unlike retailers in many other markets, water retailers have to add their margin on top of the market price for pipe-delivered water).

• While the water utility delivering water through the piped system has a social objective, and water prices are generally regulated, water vendors have no such restrictions. As long as consumers are able and willing to pay (and do not have alternatives), vendors can hike the prices. Indeed, there are several examples in the literature of collusion among vendors in order to limit competition. [27, 28, 29] In Dar es Salaam, however, prices appear to be competitive.

The utility could help address the factors (above) that contribute to excessively high vendor prices by: ensuring that more abundant, regular and reliable water is available in or nearer to currently disadvantaged area. More abundant or more easily accessible water makes the vendors' job easier and at the same time increases the competitive pressure between vendors. It also opens up for alternative provisioning routes for consumers. Offering preferential (wholesale) prices to vendors. (Such an arrangement requires some form of licensing or means of identifying vendors.)

In theory, water vendors are to have small business licences. But even though (pushcart) water-vending licences are the cheapest of all trades, vendors operate without licences. The cost in 1988 for "selling water in containers" was 150 shillings, but out of 6,082 licences issued, none was for either of these trades.[30] At the time of the 1998-9 survey, the cost of water-selling licences had gone up to T.Sh. 8,000, but remained the lowest (along with the emptying of latrines) of all trades. Together with development tax, business identity card and monthly business levies, the total annual cost for fees and taxes would amount to T.Sh. 23,900 (approx. US$ 24).[31] Given the low levels of earnings of pushcart water vendors, these obligations are practically impossible to fulfil.

Licensing or 'identification' of vendors could be done in more facilitative terms. The costs for fees and licences, if any, must be in line with potential or actual earnings of the trade. Moreover, experience from formalisation initiatives show that informal enterprises need to receive benefits of operating formally in return for paying such costs.[32] In line with this, vendors could be trained in issues of health or water quality preservation, or even to watch for leaks or the tampering with the system. According to this study, vendors appear to be interested in being associated with the utility when reaching out to customers.

There are indications, however, that rather than reaching out to already existing vendors, DAWASCO plans to make them redundant or unprofitable. This is partly through the institution of officially licensed stationary vendors (who may or may not be previously informally operating ones), but mainly by using metering as a means to curb water sales by private households.[33, 34]

Conclusions

Reliable and continuous water deliveries through a piped network are most convenient form of water service. Though it requires significant initial investments, particularly in distribution, it should remain the long-term goal for urban water development. Nonetheless, since universal coverage is out of reach in a rapidly - growing city with a limited budget – at least in the near future – it is vital for the most disadvantaged households that alternative supplies be supported rather than obstructed.

A socially responsible water policy looks at the situation of end-users of different socio-economic situations. With such a perspective, water vending cannot be ignored. In theory, DAWASCO is responsible for the whole of Dar es Salaam's water system. In practice, its role in water delivery to consumers is limited. The mainstay of water deliveries are carried out by consumers themselves as they collect water from public supplies or, most commonly, from neighbours with a piped water connection or borehole. This supply mode is the most important for low-income citizens. Those that can afford to pay others for deliveries, but lack capital to invest in a connection or live far from the distribution network, often make use of distributing water vendors. Distributing vendors in Dar es Salaam ferry water either in jerrycans on hand-driven carts or by tanker trucks. This mode is important for those with jobs – the working poor and middle-class households, as well as wealthy households and businesses whose pipe-connections are lacking or do not function well.

Vendor-delivered water, however, is very expensive. Even that collected by households themselves at neighbours or standpipes costs twice that delivered by the utility through the distribution network. This leads many households to ration their water use and consume (too) little water. The main reason for the high prices is that feet and wheels are far from being efficient for water conveyance as pipes. The main avenue for water utilities to help lower end-user prices, apart from providing piped water into all areas, is to ensure easy, reliable and abundant access to water for private re-sellers and distributing vendors.

Preferential wholesale prices to vendors, as in most other trades, should – as long as quantities are sufficient – also help lower the retailer prices to consumers. This would require some form of licensing, which ideally should be combined with training or somehow upgrading the performance of service provided by presently informal vendors.

As long as universal piped-water services are out of reach, vendors are going to be needed. Actually, given their flexibility in terms of areas served or time needed to mobilise, vendors can be regarded as a flexible extension of the piped-water system. Indeed, a five-day cut-off on the Lower Ruvu transmission line in October, 2006,[35] that was thought to completely disrupt life in Dar es Salaam instead showed the resilience of the informal distribution system. Streets were instantaneously filled with pushcart water vendors and tanker trucks, accompanied, of course, by enormous queues at boreholes and reservoirs. Most other activities, nonetheless, continued unabated even though some two - thirds of the bulk supply had been cut, demonstrating the magnitude of in-town storage and the swiftness of vendor mobilisation.[36, 37]

Notes

1. National Bureau of Statistics. 2002. *Statistical Abstract*. Dar es Salaam: National Bureau of Statistics.

2. Mujwahuzi, Mark, R. 2002. *Drawers of Water II: 30 Years of Change in Domestic Water Use & Environmental Health in East Africa*. Tanzania Country Study. London: IIED.

3. Ngware, Suleiman & Kironde, J.M. Lusugga, (eds.). 2000. *Urbanising Tanzania: Issues, Initiatives and Priorities*. Dar es Salaam: DUP LTD.

4. Strategic Urban Development Planning Framework (SUDP) 1999, draft for the City of Dar es Salaam.

5. WaterAid, "Water and Sanitation in Tanzania: An Update Based on the 2002 Population and Housing Census", available at http://www.wateraid.org/documents/2002_census_update.pdf. (Accessed on 29-03-2006).

6. Kjellén, Marianne. (2006, PhD dissertation). "From Public Pipes to Private Hands: Water Access and Distribution in Dar es Salaam, Tanzania". Department of Human Geography, Stockholm University.

7. Kjellén, Marianne, "Structural Leakage in Dar es Salaam: The Investment Deficit in Water Distribution," presented at Meeting Global Challenges in Research Cooperation, Centre for Sustainable Development, Uppsala University and Swedish University for Agricultural Sciences, 27-29 May, 2008.

8. *National Bureau of Statistics, Household Budget Survey 2000/01*. Final Report, 2002. Dar es Salaam: National Bureau of Statistics.

9. Thompson, John, et al. 2001. *Drawers of Water II. 30 Years of Change in Domestic Water Use & Environmental Health in East Africa*. London: IIED.

10. United Republic of Tanzania, The Water Laws (Miscellaneous Amendments) Act 1997.

11. PSRC, "DAWASA Hands Over Provision of Water and Sewerage Services to City Water Services": Available at http://www.psrctz.com/Press%20Releases/010803-Dawasa%20Handed%20Over.htm#topdawasahand. (Accessed 03-09-2003).

12. The Guardian,14 - 05- 2005. Bilal, Abdul-Aziz. "Govt dumps city water".

13. International Centre for Settlement of Investment Disputes (ICSID). 2007. Award. Biwater Gauff Ltd. Claimant Vs. United Republic of Tanzania, Respondent. Washington, D.C.: The World Bank Group.

14. Kjellén, Marianne. 2000. "Complementary Water Systems in Dar es Salaam, Tanzania: The Case of Water Vending," *Water Resources Development*, 16, 1, pp.143-154.

15. Kjellén, Marianne. 2000. *Uuzaji wa Maji katika Jiji la Dar es Salaam*. Stockholm: Environment and Development Studies Unit (EDSU).

16. Tanzania News Online, November 22, 1997. "Water meters soon to be installed on Dar's taps".

17. Kihinja, Hillary, a water vendor, interviewed on May 1, 2000, Dar es Salaam,

18. World Bank, Project Information Document. 2003. *Tanzania-Dar es Salaam Water Supply and Sanitation Project*. Washington, D.C.: World Bank.

19. Hodgson, G. 1988. *Economics and Institutions*. Cambridge: Polity Press.

20. Stein, Howard. 1995. "Insitutional Theories and Structural Adjustment in Africa," in Harriss, John. et al., (eds.). *The New Institutional Economics and Third World Development*. London: Routledge.

21. Harriss, John. et al. 1995. (eds.). *The New Institutional Economics and Third World Development*. London: Routledge.

22. PSRC, "Welcome to the divestiture of... Dar es Salaam Water Supply Authority", available at http://www.psrctz.com/Utilities%20&%20Major%20Transactions/dawasa.htm (Accessed on 17- 1- 2001).

23. Sykes, Adam. 1999. Small Scale Independent Providers of Water and Sanitation to the Urban Poor: A Case of Dar-es-Salaam, Tanzania. Nairobi: Water and Sanitation Program and IRC International Water and Sanitation Centre.

24. O'Leary, Mike, Chief Executive Officer, City Water Services, interviewed on March 26, 2004, Dar es Salaam.

25. Dar es Salaam Water and Sewerage Authority (DAWASA): Available at http://www.dawasa.org. (Accessed on 04 -12 - 2003).

26. Chaggu, Esnati & Edmund, John. "Ecological Sanitation Toilets in Tanzania," presented at 3rd International Conference on Integrated Environmental Management in Southern Africa, Johannesburg, August 27-30, 2002.

27. Whittington, Dale, et al. 1991. "A Study of Water Vending and Willingness to Pay for Water in Onitsha, Nigeria," *World Development*, 19, 2/3, pp. 179-198.

28. Swyngedouw, Erik. 1995. "The Contradictions of Urban Water Provision: A Study of Guayaquil, Ecuador," *Third World Planning Review*, 17, 4, pp. 387-405.

29. Lovei, Laszlo & Whittington, Dale. 1993. "Rent-extracting Behavior by Multiple Agents in the Provision of Municipal Water Supply: A Study of Jakarta, Indonesia," *Water Resources Research*, 29, p.7, 1965-1974.

30. Tripp, Aili Mari. 1997. *Changing the Rules: The Politics of Liberalization and the Urban Informal Economy in Tanzania*.Berkeley: University of California Press.

31. Komba, Human Resource Development Officer, Temeke District, interviewed on September 21, 1998, Dar es Salaam.

32. Chen, Martha Alter. 2006. "Rethinking the Informal Economy: Linkages with the Formal Economy and the Formal Regulatory Environment," in Basudeb Guha-Khasnobis, et al., (eds.). *Linking the formal and Informal Economy: Concepts and Policies*. Oxford: Oxford University Press.

33. The Guardian, Dar es Salaam, September 7, 2005. "Dawasco cracks down on water thieves".

34. The Guardian, Dar es Salaam. Kihaule, Emmanuel, August 22, 2005. "Dawasco moves to rein in water dealers".

35. The Guardian, Dar es Salaam. October 9, 2006, "Dar residents go without water".

36. Kjellén, Marianne. *"The Water Divide: Fragmented but Flexible Water Services in Dar es Salaam, Tanzania,"* presented at Association of American Geographers' Annual Meeting, 17-21 April 2007, San Fransisco.

37. Kjellén, Marianne. *"Poster Presentation, Resilient yet Unsustainable: Water Services in Dar es Salaam, Tanzania,"* presented at Resilience, Adaptation and Transformation in Turbulent Times, Stockholm Resilience Centre, April 14-17 2008, Stockholm.

Irrigation

Local Irrigation Projects in North Western Kenya, Conceptual Frameworks and Development Practices: The Missing Links

Jean Huchon & Janick Maisonhaute

Introduction

Governments and other development operators have been supporting the creation of irrigation schemes in North-Western Kenya for the last 30 years. Such well-planned projects are meant to enhance food self-sufficiency, promote economical diversification, increase rural incomes and provide employment opportunities. Irrigated agriculture is considered as a solution to a variety of problems. However, most initiatives do not achieve their goals. Some projects abort and some schemes disappear a few years after their implementation. Some persist but only with large subsidies, while others simply remain as projects.

Through the analysis of four irrigation projects in West-Pokot district, this paper addresses the gaps between conceptual frameworks and development practices. It will underline the evolution of the donors and development agencies rationales ; outline the reaction of the local populations and the real changes they induced.

Local Context and Hypotheses On Development of Irrigation Schemes

North-Western Kenya is a semi-arid region with a double rainy season with a peak of rainfall in April-May and to a lesser extent in November. During these periods, rain falls in localized connective storms and is unpredictable in space and time. Water scarcity, is a constraint to agriculture and it is therefore not surprising that livestock husbandry is the main economic activity for the semi-nomadic communities inhabiting the Rift plains, the Turkana and the Pokot. Nevertheless, some populations are used to practicing traditional agriculture.

The Turkana Cultivation System

The Turkana are well known as pastoralists and do not have a strong cultural background in agriculture. However, the Ngebotok Turkana, of the middle Turkwell, enjoying more favourable conditions, rely more heavily on agriculture than other Turkana sections: *"Probably the simplest practice of floodplain cultivation recorded in East Africa is that of the Turkana in Kenya, who until very recently cultivated sorghum in the floodplains of the Kerio and Turkwell Rivers. Gardens were laid out on meander scar and terrace land, and cropped in the rains, notably at*

Nagaloki on the Kerio, at Kaputir on the Turkwell, and where streams made the shore of Lake Turkana cultivable, in the Kerio River Delta and at Kalokol at Ferguson's Gulf. Cultivation appears to have been an integral part of the primarily pastoral subsistence strategy. Competition with formal irrigation development, and the social and economical implications of recent droughts and associated relief responses, such as formal rainwater harvesting schemes, appear to have substantially reduced this cultivation".[1]

The Pokot Irrigation System

According to B.E. Kipkorir,[2] agro-pastoralism was the first economic activity used by the Pokot on the Cherangani Escarpment, before they moved to the eastern plains at the beginning of the 19th, then to the western plains and became pastoralists like the Turkana and the Karimojong. There is no data regarding the construction of the first irrigation furrows in the district, but oral traditions suggest that it occurred hundreds of years ago.[3]

The furrows are concentrated in and around the Cherangany Hills and the Sekerr Hills. *"Where the terrain and soil cover permit, the furrow is excavated out of the slope, the original tools for this, according to tradition, being wooden digging sticks. The lower bank requires reinforcement against the erosive scour of the water and this is usually done with whatever stone is available. The ideal is a neat lining of flat stone slabs but such convenient material is rarely available. The stonework is consolidated by encouraging the growth of vegetation –tree or shrub roots in forested area or, where sufficient light penetrates, grass and the large sedge, yashan (cyperus alternifolia). The width of the furrows in this sort of situation is rarely more than about 120 cm and usually around 60 or 70 cm. Velocity of flow is usually between half a metre and one metre per second but may be up to two metres per second if the channel is constricted. On the valley floor, stone is very rarely available for lining and erosion may be a problem, though gradients and hence velocity are of course less."*[4]

In fact, the Pokot know two types of fields: the rain-fed ones and the irrigated ones. Each has its own value in the various eco-zones. The former, situated in the highlands, need a longer period to mature. The latter produces the crops more quickly. The maintenance of such an irrigation system *"needs the cooperation of all people concerned. Irrigation canals sometimes carry the water over a distance of several kilometres. Through furrows and branches the water reaches the individual plots where it floods the field. Strong regulations are heeded which give everyone his share of water and maintenance; a sophisticated organization and often a bone of contention. Through this relatively advanced technique - compared with other agricultural peoples in Kenya - people are able to make use of several eco-zones and thus reduce the risks of drought and disease."*[5]

Traditionally used to irrigate sorghum crops when rainfall failed during the growing season, the use of furrows for irrigation has decreased since the beginning of the '60s. The introduction of maize -with larger yields, easier husbandry- has caused a considerable decline in sorghum and millet production, thus making irrigation less necessary.

Pastoralism itself is also affected by several constraints ranking from low productivity of the rangeland and the low, patchy and unpredictable rainfall. Pastoralists's livelihood is thus extremely insecure. For the most pessimistic studies, the whole production system is now collapsing due to a desertification process associated to the rise of inter-ethnic conflicts over increasingly scarce natural resources. In 1974, 1979, 1984, and more recently in 1999-2000 as a result of severe droughts, a number of pastoralists lost their herds and were forced to flee to small "buying centres" waiting for food relief. Others raided their neighbouring tribes to recover animals, built a new herd and returned to pastoral life. However, they were themselves under the threat of raids and most of them were unable to rebuild their cattle herd.

Conceptual Framework of Integrated Modern Irrigation Projects

Established in this context of extinction of traditional forms of agriculture and pastoralism under severe pressure, the irrigation projects had diverse objectives. On the one hand, the initial aim was to provide a source of food and livelihood for impoverished pastoralists, war refugees and internally displaced people. On the other hand, donors hoped to promote economic development of the region through agricultural marketing and hydropower.

The irrigation schemes were part of a broader plan. For the development operators, the common idea was that the settled pastoralists would spontaneously and definitely change their way of life to agriculture specialisation. They thought that food crop development and livestock development were closely related, and when a farmer had sufficient food, his reliance on livestock reduced. Then, these settled and de-stocked communities would farm and produce enough to no longer depend on food relief, with surplus to be sold to neighbouring famine prone areas. In the long term, they would supply other pastoralists living in the countryside and specialized in better management of livestock activities. The livestock being considered as the main cause of the environmental degradation, the reduction of livestock populations and the new management of the livestock activities in the countryside would have an effect on the preservation of the environment.

The irrigation agriculture projects were also supported for security reasons. By offering employment opportunities to the youth, who constitute the main raiders, it was seen as a preventive measure to reduce violent conflict. Moreover, from a military point of view, irrigation schemes would settle the pastoralists and create a buffer zone between the pastoral groups in conflict: zone with high density of populations would constitute a border between the less populated pastoralists conflicting areas. Farmers from diverse ethnic groups associated to the management of the irrigation schemes would constitute a multi-ethnic community working as a cultural bridge between hostile groups. A market place would also bring people to exchange products and develop social relationships in a peaceful environment.

According to this argument, irrigation seemed to be an exceptional integrated response to what some people still call the decline of pastoralism. However, the conceptual framework is not always realistic, and a number of constraints reduce the chance of success of such ambitious projects.

Four Irrigation Schemes Along the Turkana/Pokot Border

From 1975 to 2005, four irrigation scheme projects were launched successively and independently. Each has its own history, its own special stakes and does not have any links with another project, in spite of geographical proximity.

The Amolem Irrigation Scheme (1975-1985)

Located on the Turkana and West-Pokot district border land, the Amolem irrigation scheme was established in 1975 on the eastern bank of the Wei Wei River, about 5 km off the tarmac road from Kitale to Lodwar. The initial project was financed by the Kenyan government, with support from the Food and Agriculture Organisation (FAO), the United Nations Development Program (UNDP), the Norwegian Agency for International Development (NORAD), the Catholic Church and the Reformed Church of East Africa (RCEA). The Amolem experimental development project was part of a broader program, which planned to create several irrigation schemes in north Kenya: Kekarongole and Katilu (Turkana district), Madera (Mandera district), Mertie (Isiolo district), Mbalambala (Garissa district), Malka Dakaa, Gafarsa, Amolem.

These schemes were designed with the objective of settling nomads affected by occasional droughts, thus providing them with an alternative livelihood. Irrigated agriculture was expected to augment food security and reduce food relief supply. In spite of their ambitions, the Amolem irrigation scheme disappeared in 1985 because of the complexity of the management, the absence of any form of local population participation and especially due to the withdrawal of support from donors.

At the beginning of the '70s, the site of Amolem was un-inhabited. The population left the area due to the Pokot and Turkana raids and because they suffered from the effect of shortage of food. The first phase of the project organized the scheme especially for these returnees. Both Turkana and Pokot were allowed to have one plot on the scheme, with a system of ethnic quota for land allocation. The aim of the managers was to re-populate the plain and establish a green buffer zone, which would help and even stop the raiders movements. They also wanted to create a local community development project tied to the churches and education through schools. The Amolem mixed community was supposed to establish a socio-economical bridge to improve and reinforce peaceful relationships between the two hostile groups. In fact, *"the community associated with the Amolem project in 1975 grew to 1,500 people in a few years' time. Among them many Turkana were allotted plots, but these were refused by Pokot people".*[6]

Initially, irrigation water was pumped from the Wei Wei River to the primary canal, while application to the crops was by ridge and furrow. This system necessitated considerable machinery and fuel. During the first years, an intake canal of 2.7 km was dug to allow the water to reach the scheme through gravity, but the transit loss of water was more than 50%, the pumps were not used for long periods and the fields further away were not used. To solve these technical problems, in 1983, it was decided that the method of irrigation of the main field would be changed to basin irrigation with gravity-fed water supply.

Notwithstanding this technical modification, only 12 hectares out of the 52 hectares of the scheme were cultivated. The production of maize and sorghum was never enough: from 804 Kg/ha of maize in 1980 to 2000-3000 Kg/ha in 1984. Even during the good years, only a few farmers could sustain themselves. The population of the scheme was still dependant on food relief, which was not an incentive for them to give more attention to their plot. The managers organized some "food-for-work" projects to motivate the population to farm or to maintain the irrigation furrows, but success was limited. Work on the scheme quickly became associated with food payments, and when annual repairs were required, labour could not be recruited until food payment was arranged. The use of food payments as a compensation for labour jeopardized the sustainability of local initiatives after the withdrawal of food aid.

In fact, the tenants were employees in the schemes rather than full time farmers. There were no farmer associations or local management committees. The participation of the population was minimal. Every season, every year, the farms changed from a family to another, and the monitoring was not done. The farms were systematically attributed to men though most of them left the place and their family soon after to look for a job in the administrative centres or as herders for some richer families living in the countryside. Consequently, the population was composed of a majority of mothers and their children, which made them lose their farms when their husbands were not back for the harvests.

The only successful farmers on the scheme were not destitute pastoralists, but farmers who came from Sigor or Orwa where they were already involved in agriculture[7]. Under their pressure, a few years after the beginning of the scheme, the managers tried to encourage the production of other crops, such as cotton, green grams, sunflower, cowpeas, simsim, groundnuts, banana, citrus fruits and vegetables. The administration believed that this solution was the best strategy to the financial woes of the project. But, there were no market opportunities to sell an eventual surplus.

Added to the agricultural problems, the preservation of the environment was an issue. The process of desertification was exacerbated rather than alleviated by the scheme. The population was growing rapidly. In 1984, 220 families were living in Amolem; this number was superior to the population of most major administrative and trading centres of the West-Pokot and Turkana districts. The population was especially concentrated around the scheme, cutting the trees to get cooking fuel, or burning charcoal to sell on the Lodwar-Kitale road. Because of this, thorn scrubs colonized large areas, dominating more useful species, in the uninhabited pastoral plain of Masol and the population was accused of being mainly responsible for environmental degradation. Moreover, irrigation development had made little to provide an alternative to pastoralism. Agriculture was just a way to build a new herd and return to pastoral life. Farmers invested in small livestock and sent them to pasture in the periphery of the schemes. Therefore, overgrazing destroyed the vegetation surface and contributed to increase erosion around the scheme.

Finally, from the point of view of the supporting organizations, the cost of these irrigation schemes was huge. R. Hogg[8] explained that the 1984 Ministry of Agriculture evaluation estimated total expenditure in 1983/1984 at US $61,240 per hectare or $21,800 per tenant household, and the operating costs alone amounted to over three times the gross margin any farmer could expect from his plot. Faced with this financial evaluation, in 1984, the UNDP and the FAO stopped sponsoring the project while the Catholic Church attempted to manage it alone. But in 1985, the project stopped all together. In the same year, the building of the Turkwell dam began some 20 km from Amolem. Despite the failure of the Amolem irrigation scheme, the initial Turkwell project planned to establish a new irrigation scheme based on the same mode.

The Turkwell Irrigation Scheme Proposal (1987-2005)

Internationally conceived, the Turkwell Gorge Project is considered as one of the biggest white elephants projects of the '80s. In fact, it was first conceived in the late 1960s by the United Nations Food and Agriculture Organization (FAO) and Norwegian Agency for International Development (NORAD) in one of the most spectacular sites of north-western Kenya.

Environmentally, the Turkwell gorge area is known as *"very sensitive to both nature and manmade conditions. In drought the hardy and thorny acacia stands alone on patched dry and exposed ground while during the rains the undergrowth is rich with thick growing grasses and other leafy shrubs. Slightest disturbance by man through cultivation of hill sides and river valleys, intensified livestock grazing and bush clearing for charcoal burning or extension of farmland has led to untold losses of the surface soil with the consequent danger of desertification. This has drastically reduced the land's potential carrying capacity [...]".[9]*

The initial aim was multi-purpose with two major components: to provide hydroelectric power and to create an irrigation scheme downstream. The first assessments were undertaken by consulting firms from Sweden and Norway. Further studies were carried out by a Norwegian consulting firm. The last report, issued in 1976, concluded that even a single-purpose dam for energy production alone could be a favourable undertaking in economic terms. However, during the 1979 State visit in France, the Turkwell Gorge Project was discussed between the Kenya and the French government representatives. Following this discussion, it was agreed that the French government would allocate a grant to finance the preliminary design of the project. A French consulting firm was commissioned for this purpose and submitted its preliminary report in 1983 which recommended the construction of a double curvature concrete arch dam at the entry of the gorge to provide storage volume of 1,080 million cubic meters, available for the irrigation components.

Finally, the hydropower project with an installed capacity of 106 Mw of electric energy, connected to the national grid, through a 230 km long power transmission line, was commissioned in 1991. But, the irrigation scheme downstream component was cancelled from the global project. According to the consulting engineers who were in charge of the design and construction of the project, the implementation of the 1,286 hectares identified as irrigable land and the design of a pilot scheme of 405 hectares had been hampered by lack of funds.

From 1990, the Kerio Valley Development Authority (KVDA), the parastatal operator, tried to push for the renewal of the second component of the project, denouncing the indirect effects of the dam to explain the necessity to implement the irrigation project. On one hand, it argued that *"the change in the river flow regime as a result of hydropower development has resulted in the degradation of the riverbed downstream of the dam, with attendant lowering of water level causing destruction of the riverine vegetation which adversely affects the population downstream. Therefore mitigation measures needed to be put in place to counteract the effect of riverbed degradation"*. On the other hand, it supported the need for compensation dispositions, arguing that *"The project area is semi-arid, remote and the inhabitants are pastoralists and do not benefit directly from the hydropower development. [...] The irrigation component which has not been implemented was meant to benefit the local populace in ensuring food security and hence minimize reliance on food relief and also in changing lifestyle. It is also clear that irrigation development will help in integrating the Pokot and Turkana communities and therefore cattle rustling for economic reason which is evident today will be controlled [...]"*.[10] Considering that at the beginning of the dam project, the effects of the dam would be insignificant and that *the problem of resettlement of the displaced population* would not arise because there were no permanent settlements or agricultural activities within the reservoir area", the KVDA estimated some years later that the future of the same pastoralists affected by the dam was in irrigated agriculture. In reality, the irrigation scheme would be a great source of revenue for organizations such as KVDA.

In response, the donors claimed that such an irrigation project would be too expensive because of the cost of pumping water. Indeed, the electrification of the area had not even been done, because the engineers thought that the local population, as pastoralists and not completely integrated into the monetary economy, would not pay for services (less than 7% of the West Pokot District was served with electricity). Nevertheless, electrification of the area would reduce the cost of the irrigation infrastructure and would improve the integration and economic development of the area, which could support the cost of the electrification investment. In other words, although talking about long term investments, the donors seemed to be looking for short term profits.

In the meantime, the dam had adverse socio-economic impact on the local population, which had not been prevented nor mitigated. The Pokot leaders explained that the flooding of the Suam River had deprived them of their most

191

traditional meeting places which were the focus of their judicial and political activities, as well as their recreational and work places. The lake had limited their ability to communicate easily with people across the river. It had in some cases separated families and it had become difficult for people to cross, with their livestock, to the other side of the River. In addition, the presence of crocodiles in the lake posed a real danger to the peoples' livestock. Finally, the dam was fertile breeding ground for mosquitoes and in the absence of health facilities, malaria had become a serious problem. In 2006, the Pokot were still complaining that the reality was the exact opposite of what they had been promised.

The Wei Wei Commercial Estate (1987-2005)

The last irrigation scheme example draws inspiration from the success of Wei Wei irrigation scheme. This irrigation scheme was implemented in 1987 by the Kenyan government, through KVDA and the Italian Development Co-operation. It was a part of the Wei Wei Integrated Development Project (WWIP). The project planned the development of an area of approximately 700 hectares, with 580 hectares being equipped with gravity-fed overhead irrigation. The irrigation system supplied water without any restriction. In 2005, 300 hectares out of the target 700 hectares were allocated to tenant farmers. A total of 225 plots were divided into 2.5 acre plots. Each plot had a hydrant fed from an underground pipe network and was equipped with galvanized steel irrigation laterals, raisers and sprinklers.

The initial aim of the scheme was to deal with food shortage and assure the food self-sufficiency of the population in Sigor, where many Pokot pastoralists took refuge after the Turkana raids on the Masol plain during the '70s. In fact, in 2005, the majority of the tenants were not refugees or destitute pastoralists, but Pokot from the hills. These populations were already experienced in growing crops, having used their traditional furrow irrigation system for a long time. Therefore, the aim of the scheme re-focused on economical development and market integration at the expense of curbing food shortage in the area.

For this last reason and because the price of food crops was low and the crop budgets showed that farmers could never afford to pay for inputs (chemicals and fertilizers) and services imposed, a short time after the creation of the scheme, the administration decided to abandon its food grain policy in favour of onions and peppers, and later, hybrid maize seeds for the export market. Many arguments supported this market orientation: (1) the location of the scheme had an important potential for this kind of dry production; (2) contrary to fresh vegetable produce, spoilt by transportation, dry productions benefitted from good climatic conditions, which helped quick dehydration of the crops (3) moreover, the Wei Wei farmers could sell their seed productions to the three main Kenyan seed companies (Kenya Seed Company, Western Seed Company and East African Company), at a much higher price than the market level and the farmers would make a good profit.

This new orientation was in consultation with the Wei Wei Farmers' Association.[11] Therefore, in 2005, the Wei Wei irrigation scheme made a significant contribution to employment and income generation in the area, stimulating the development of economical transactions in the small centres. This was illustrated by the expansion of the Sigor market. The example given by the tenants tended to be imitated by other farmers. Many households not involved in the project cleared pieces of land next to the project. They hired pipes and sprinklers and drew water from the plots of farmers who were beneficiaries. In the same way, some communities from the Masol plain and from the Sekerr foothills adopted irrigation schemes for the creation of a market turned towards agriculture. People living around the irrigation scheme benefited from the constant supply of water both for themselves and for their animals. The Wei Wei irrigation scheme was considered a success by most of the locals.

Rehabilitation of the Orwa Irrigation Scheme

The Orwa irrigation scheme revival is one example of an irrigated agriculture extension inspired by the Wei Wei irrigation scheme success. Located in the West-Pokot district, some kilometres from the Turkana district administration border, this indigenous irrigation scheme had been abandoned in 1999 because of the internal Turkana-Pokot raiding. Former farmers came back in 2004 after the installation of a police station nearby. The entire infrastructure was destroyed. It was in this context that an international NGO planned the rehabilitation of the scheme. This project was a part of a larger project dealing with Pokot/Turkana conflict resolution. Schools and irrigation schemes projects were planned on both sides. The project was engaged on the Pokot side in Orwa in 2005.

Before 1997, approximately 200 households farmed in Orwa, while the total surface area of the scheme was about 250 hectares. The water was diverted from the Orwa River with a dam of wooden poles and stones at the foot of the Sekerr escarpment, and then conveyed through an earthen and rocky furrows system to the fields. Three main furrows divided the schemes into three distinct zones. Up to the upstream, a first furrow irrigated a reduced field where each farmer cultivated sorghum on 0.25 hectares. Lower downstream, two other main furrows were able to irrigate some larger plots, from 1 to 2 hectares for each household, through several feeder furrows. These small secondary furrows went along the edge of the individual's farm rather than through the middle of it, serving as boundaries between the plots.

Irrigation enabled two harvests to be produced per year. The first one depended mainly on precipitation from March to July. The furrows were opened only when the rain was not sufficient at the end of the season. The second harvest depended entirely on irrigation water supply. The small secondary furrow maintenance was usually carried out at the same time as the farmers were watering their plots. The maintenance of the main infrastructures, intake as well as main furrows, and the distribution of irrigation water was organised by the *kokwö* council which gathered all the heads of households involved in irrigation[12]. The *kokwö* had to oversee and co-ordinate the annual preparation and maintenance of the main furrows. Members who neither contributed with their own labour nor hired anyone to do the work were fined the average agricultural wage rate for days not worked. In theory, during irrigation, the use of water to the farms had to be regulated. In practice the committee or *kokwö* did very little once the task of reopening main irrigation canals was completed. Producers usually fenced and cultivated according to their own schedules and in agreement with their neighbours. When conflicts arose, the *kokwö* served as intermediaries for disputing parties. However, local conflicts were rare. Irrigation agriculture being labour intensive, it was necessary to keep good neighbourly relations. Co-operation was often beneficial as it enabled farmers to mobilize extra labour to prepare their fields in time, to devote more attention to agriculture during the late dry season when the herding of cattle restricts the amount of labour that can be allocated to the farms.

Contrary to the organization of the indigenous scheme located on the Cherangani escarpment, which is based on lineage and clan ties, and in accordance with land pressure, the Orwa type of organization was essentially based on strong neighbourly relationships. This system of management was close to the pastoral management system. It allowed more flexibility, probably in response to the risk of the Turkana raid in the area. Actually, the relationship between farmers from the schemes and herders was relatively important, even with the Turkana herders during peacetime. The pastoralists, on the way to the dry season pastures, came to exchange cereals for animal products: meat, blood and milk, which were bartered for sorghum and maize. Most of the farmers from Sekerr Hills as well came to barter or to sell their produce. Relations between these different communities were as strong as inter-ethnic marriages between Turkana and Pokot. Orwa households left their animals with pastoralists, or some Orwa farmers had farms on both sides of the escarpment, in the Sekerr Hills and on the scheme. A weekly market was even organized to favour these exchanges.

In spite of these relations, the Turkana raiders attacked the market in 1997 and in 1999. After revenge from the Pokot, the place was abandoned and became a no man's land. The conflict between Turkana and Pokot continued through 2006. NGOs wanted to revive the irrigation scheme and the market to reinstate relations between the two foes. It considered the irrigation scheme as a protection and a conflict resolution tool. Except for the technical problems, the first stage of the rehabilitation project was to deal with the return of the refugees and the organisation of a water committee to implement the procedure of land allocation and good governance with the objective of protection and avoiding potential internal conflicts. Indeed, the refugees were scattered, living on the periphery of the small urban centres of Marich and Sigor, or in the Sekerr Hills where they were welcomed and they cultivated the slopes of the mountain.

Moreover, the first farmers who returned to the scheme were mainly young men. They became members of the newly formed water committee, while in the past, elders dominated the *kokwö* in charge of the management of the scheme. They adopted the government irrigation committee model through the District Social Development Officer as a Registration of Help water committee group. This model, which required eight committee members: Chairperson, Vice-Chairperson, Secretary, Treasurer, and four non- office holders, was a requirement for government financed small-scale irrigation schemes. The presence of educated men on the committee, sometimes local politicians and/ or government employees, probably accounts for the use of the official formula. These individuals were familiar with administrative procedures and were able to eventually obtain external assistance for the scheme. Indeed, the committee positions assumed considerable importance for aspiring leaders, who used them to mobilize political support. At the least, this was a good strategy to build real cooperation between the generations, the youth from the water committee with the elders of the *kokwö* and keep the community focused on the irrigation scheme.

The Impact of the Irrigation Schemes

Previously, the irrigation schemes were considered exclusively and without participation and interaction from the communities. The efficiency of the irrigation schemes was not only a matter of productivity, technology or a model of farmer organisation. A scheme has to be accepted and "appropriated " by the local populations, not only by the farmers directly involved but also by the other communities living nearby. This is the main condition to enable and guarantee a certain long-term viability of the scheme. Indeed, an irrigation scheme cannot be considered without any indirect socio-economic effects on the local society. Pastoralism involves a specific spatial economy linked to the herders' transhumance patterns and the effects of an irrigation scheme in a pastoral area are consequently important, despite not being so obvious. An irrigation scheme has definite socio-economic impact on the local society and generally affects the political stakes.

First, it is important to keep in mind that agriculture is not independent, but complementary to other production activities. Although livestock husbandry is the major economic activity of the Turkana and the Pokot populations who inhabit the Rift Plains, agriculture, and especially irrigation agriculture, is not an innovation. Some communities living on the Cherangani foothills were using a complex irrigation system with furrows built over five hundred years ago. In fact, agriculture is such an integrative activity for pastoralism, that agricultural production is a condition for the viability of pastoralism; the pastoralists, to avoid affecting their herd productivity which could endanger the food requirements of the household, need to compliment their activities with agricultural production. When it is possible, they produce cereals themselves or buy them from cereal producers. Water scarcity being the main constraint on agriculture, the irrigation schemes are the main agricultural production places and constitute strategic areas. But, the case studies also show that irrigation programmes are more viable when dealing with populations which are experienced in irrigation and use their own models of management rather than with destitute pastoralists.

Irrigation scheme populations are also linked to pastoralism, which offers socio-economic empowerment opportunities. Livestock attribute such social, economic and political prestige that every household maintains a preference for pastoralism over agriculture[13]. Without any cattle, a man no longer needs to move from one place to another, therefore he cannot extend his relationship network, which is necessary to get prestige and recognition among his own community.[14] Moreover, the pastoralist's economic superiority over the agriculturist's is presented in the way that the pastoralist gains a real commercial profit when he exchanges cereals for their animal products. Agriculture is part of a survival strategy for pastoralists who have lost their herd after a crisis (drought, raiding). Due to this, they revert to temporary farming of cereals, with the hope of returning to their pastoral life later on. Indeed, the reinvestment of the agriculture surplus in livestock enables them to recover and have a new herd after some years.

According to Anderson's statement: "*To be poor for the pastoralists means that because of the loss of their wealthy livestock, they are no more pastoralist, but simply 'poor' agriculturists*".[15] "*Puryö tïch, kima mren*", a man without any cattle is a dead man, says the Pokot proverb. In this context, it is problematic to consider irrigation success as proof of economic development. The irrigation scheme projects which develop the arguments of fighting against poverty and plan to "make" the pastoralists into agriculturists contradict this latter statement. This can be, to a great extent, be explained by the failure of the Amolem irrigation scheme; the administration based the successful outcome on an experiment on productivity, while a destitute pastoralist based the success of the irrigated agriculture on his capacity to return to a pastoral way of life. According to this interpretation, when dealing with destitute pastoralists, the disappearance of irrigation schemes like Amolem can even be considered as a success.

Conclusion

There is a real misunderstanding between the aspiration of the destitute pastoralists and the models proposed by the donors and national authorities. On the one hand, no homestead expresses the desire to become full-time farmers; pastoralists view irrigated agriculture as a stage, thus irrigation remains a supplementary activity, with communities and homesteads preferring flexibility in order to diversify or to accommodate other economic activities, especially pastoralism. On the other hand, donor organizations consider irrigation projects as a long-term investment that will offer quick profits. Parastatals, particularly the KVDA or National Irrigation Board (NIB), consider irrigation as a source of income, whereby crops, having a comparative advantage at the national level, can be grown. In spite of the lack of local market, crops like onions and peppers are promoted, while commodities with a strong local demand, such as maize and millet, are discouraged.

In actual fact, the latter analysis on transformation from pastoralism to a more agricultural way of life is particularly effective in remote areas where the export market is not efficient. In contrast, once the market is accessible, irrigated agriculture becomes the motor for integration, implying a deep change of the socio-spatial dynamics of the society. When production is oriented towards the market channels, irrigation and cattle rearing activities tend to be pursued as a way of diversifying income-earning opportunities. As a result, a strong correlation exists between the size of herds and the size of irrigated farms.

Both livestock and irrigated land are increasingly being monopolized by rich homesteads which exaggerates wealth differences, enhancing the position of the rich herder *vis-à-vis* the poor. Indeed, rich livestock owners control the necessary capital and labour to meet the high costs of irrigation and are able to cultivate the largest farms; this enables them to build up livestock herds by keeping market sales at a manageable level. They participate in the upward cycle of increased grain production and increased livestock holding, while the poorer herd owners are caught in the downward cycle of de-capitalisation and are compelled to sell their animals in order to meet subsistence costs.

These socio-economic changes can be analyzed on a spatial level. In particular, the process of capitalization implies changes in land tenure, which could promote conflict. In most of the schemes, the richest agro-pastoralists cultivate the best irrigated plots. In spite of land rights governed as land trusts, they fence their farms to secure them. These fences symbolize the "land competition" context as an effect of the commercialization of agriculture, which also increases land value. Consequently, the poorer agro-pastoralists practicing dry land agriculture or low-cost forms of irrigated agriculture has no other solution than to extend farms on the periphery of the schemes. This expansion is increasing as the number of impoverished pastoralists increases. Paradoxically, the impoverishment of the pastoralists is also an effect of the extension of the scheme on pasture land, which interferes with herd movements consequently affects herd productivity.

Usually, pastoralists are reluctant to propose a ban on farming because they themselves also practice small-scale agriculture. Indeed, the cattle herders sense the danger of losing land when the peasants plant permanent crops or develop their farms by erecting permanent structures. As a reaction, some of them tried to farm maize and beans near the river as an attempt to mark the landscape and to claim permanent occupancy of the land, which can be used as grazing land after crop harvest.

In spite of these internal rich/poor differences, the threat of land appropriation by other groups is strong enough at times to mobilize the local communities against outsiders. Because of the regional competition for land among ethnic groups on the Kenyan Highlands, the original homeland of Pokot politicians, cultivation becomes a way to maintain a claim on the land, even if economic returns are minimal. In fact, agricultural investments can be seen more as a method of securing access to territorial land, than as a strategy for increasing short term profit. This can explain why the politicians are not supporting, as much as they could, the Turkwell irrigation scheme located on the Turkana/Pokot Border. Such a project would institute consensus on the border and as a result none of these ethnic groups would be able to continue to claim the land as their ancestral home area.

Most of the irrigation schemes have been implemented, or at least planned, as independent local development projects without any regional approach. In all cases, the first objectives have been diverted from a local food supply to an export market supply, whereby only crops having a comparative advantage at the national level could be grown. Consequently, and in spite of a certain success, the production of export crops, red peppers, onions and vegetables, has not contributed to increase the amount of surplus grain available for trade within the local areas. The study area is today still under the recurrent food deficit menace. In other words, the process of integration, which is conceived as a development process, is proceeding, but famine still persists.

The irrigation projects have also not contributed to the resolution of regional conflicts. Orwa scheme was abandoned again because of conflicts. In future, it would be important to reconcile irrigation with a set of productive local relations. This is dependent on political will, which is sometimes more concerned with national and economical stakes than with local development. Finally, the experience of these projects reminds us that a local project has to simultaneously integrate local, regional and national - level components.

Notes

1. Adams, William M. & Anderson, David, M. 1988. "Irrigation Before Development: Indigenous and Induced Change in Agricultural Water Management in East Africa." *African Affairs.* *Vol. 87*, p.523.

2. Kipkorir, B.E., Soper, R.C., & Ssennyonga, J.W. 1983. *"Kerio Valley: Past, Present, and Future".* Nairobi : Institute of African Studies, University of Nairobi. p. 171.

3. Klinken, M.K. Van,.1987. *"The Pokot Traditional Furrow Irrigation: The Construction of Aqueducts in West Pokot District, Some Data and Experiences",* West Pokot District Information and Documentation Center, Kapenguria, Report for ASAL Programme, West Pokot. *"In the semi-arid areas of Eastern West Pokot 'traditional' forms of gravity fed furrow irrigation existed long before colonial times, albeit on a very small scale. They were already reported by Austin in 1903, and they were gradually expanded during this century".*

4. Soper, Richard. "A Survey of the Irrigation Systems of the Marakwet in Kerio valley, in Kipkorir, B.E., Soper, R.C. & Ssennyonga, J.W. 1983. *"Kerio Valley: Past, Present, and Future".* Nairobi : Institute of African Studies, University of Nairobi. pp. 8-9.

5. Visser, J.J. 1989. *Pökoot Religion: Oegstgeest".* Hendrik Kraemer Institut. pp. 12, 13. This irrigation system is a part of the larger Kerio cluster furrow systems of East Africa, which *"is found on the Marakwet escarpment and the Wei-Wei Valley below the Cherangani Plateau in Kenya. Marakwet and Pokot villages have superficially similar furrow systems diverting water from streams descending the escarpment and carrying it in earth and sometime stone channels down the escarpment face to irrigate substantial areas on the valley floor. These furrows show considerable engineering skills and represent what is probably the most extensive and complex indigenous water management in Africa, south of the Sahara. The escarpment along this east wall of the Rift Valley is spectacularly precipitous, with furrows cutting into the escarpment streams at altitudes of 9000 feet and dropping to the valley floor at 3000 feet. The bulk of cultivation of cereals and vegetables take place down in the valley between the foot of the escarpment and the east bank of the Kerio Valley, but the irrigation furrows also feed smaller "kitchen gardens" and some larger fields of cereals among the villages on the hillside. In places these cultivations are*

supported by elaborated systems of terracing. The history of irrigation in this area has not been thoroughly investigated but it is clear that the furrow system was as extensive in the nineteenth century as it is now and it is possible to chart the decline in the use of older canals and the construction of new furrows during the present century". Kipkorir, B.E., Soper, R.C. & Ssennyonga, J.W. 1983. "Kerio Valley: Past, Present, and Future". Nairobi : Institute of African Studies, University of Nairobi. pp. 1-22.

6.   Visser, J.J. 1989. Pökoot Religion: Oegstgeest". Hendrik Kraemer Institut. pp. 49.

7.   Dubel/De Kwaasteniet. 1983. p. 69., Visser, J.J.1989. p. 50. A conclusion from a study about this project stated as follows: "Past policies have shown however that firstly, projects are doomed to fail when the people concerned do not agree with them and secondly, that the people from Sangat / Weiwei accept changes very slowly".

8.   Hogg, Richard. 1988. "Changing Perceptions of Pastoral Development : A Case Study from Turkana District", in Brokenska, DW, & Little, P. D. Anthropology of Development and Change in East Africa. London: Westview Press. pp. 183-201. Little, PD. 1992 The Elusive Granary, Herder, Farmer and State in Northern Kenya. Cambridge: University Press. p.212.Maisonhaute, Jannick. (2002, PhD dissertation) "Jouer avec le Paradoxe: L'exemple des Pokot du Kenya", University of Paris X.

9.   Were, F.B.K. 1983. "Responsibilities and Activities of the Kerio Valley Development Authority", in Kipkorir, B. E. Soper, R.C. & Ssennyonga, J.W. "Kerio Valley: Past, Present, and Future". Nairobi: Institute of African Studies, University of Nairobi. p. 1.

10.   Internal French Embassy archives. 1985. "Letter from the Kerio Valley Development Authorities Director to the French Embassy" ASAL Programme. p. 41. The Kerio Valley Development Authority was established by the Act of parliament N° 14 of 31ˢᵗ August 1979. This act was initiated with the sole purpose of trying to focus interest and resources into this area which for historical, geographical and other reasons has lagged behind in general socio-economic development within the Kenya context. In consequence therefore, the Authority has been charged with the following functions. To initiate, identify, plan and implement feasible development projects and programmes in consultation and collaboration with the speciality Government Sectoral Departments, local leaders and administrative officers... To develop and keep an up-to-date record of short, medium and long range development plans for the area... To initiate and undertake reconnaissance studies, speciality surveys and any other pertinent exercises which may be deemed necessary pre-requisites for more detailed feasibility studies, or pre-investment studies and engineering design." Were, F.B.K. "Responsibilities and activities of the Kerio Valley Development Authority". Kipkorir, B. E., Soper, R.C. & Ssennyonga, J.W. 1983. "Kerio Valley: Past, Present, and Future". Nairobi : Institute of African Studies, University of Nairobi, p. 1.

11.   An association especially created by the direction at the beginning of the project in 1991 to promote farming activities in the area.

12.   Visser, J.J. 1989. Pökoot Religion. Oegstgeest". Hendrik Kraemer Institut, "Every neighbourhood (korok) has its own kokwö, which may mean council as well as a meeting place. A man of any clan and age-set can participate. In these meetings people discuss matters of common interest such as communal labour, as in the case of an irrigation channel, relations between the neighbourhood, problems of security, etc. The councils also uphold law and order and thus actually function as a court in any dispute or case. The power - political and judicial - is vested in these bodies of adult men, the elders, the poy".p. 76.

13.   Herskovits. 1926. "Calls this 'Cattle complex', an intense devotion to cattle and permeation in the value into all aspects of culture".

14.   Maisonhaute, (2002, PhD dissertation). Jouer avec le Paradoxe: L'exemple des Pokot du Kenya. University of Paris X, p.592. Huchon. (2004, PhD dissertation). Intégration et Conflits d'espace (s): Les Dynamiques Territoriales des Pokot, Nord-ouest du Kenya. University of Toulouse Le Mirail, p. 420.

15.   Adams, William, M. & Anderson, David, M. 1988. "Irrigation Before Development: Indigenous and Induced Change in Agricultural Water Management in East Africa." African Affairs. Vol. 87, p. 819.

Further Notes

- Dietz, T. 1987. Pastoralists in Dire Straits : Survival Strategies, *Institut Voor Sociale Geographie*, University van Amsterdam. p. 323.
- Hogg, R. 1994. *Pokot Traditional Irrigation and its Future Development*. Nairobi: The World Bank.
- Hubert, H. (ed.).1985. *District Atlas: West Pokot*, Kapenguria: Ministry of Planning and National Development, Republic of Kenya. p. 136.
- Rambaldi, G. 1992. "The Wei-Wei Integrated Development Project, Small-scale Irrigation in Remote areas: An Approach to Marketing as Contribution to Successful Project Implementation", presented at the workshop on Experience and Perspective of Horticultural Small-holder Irrigation Schemes in Nairobi, Kenya.
- Zaal, F., Van Tienhoven, I. & Schomaker, M. 1985. *Masol Location, West Pokot District, Kapenguria : Arid and Semi-Arid Land Programme*.

A Happy Marriage of Traditional and Modern Knowledge - Shallow Wells:
A Sustainable and Inexpensive Alternative to Boreholes in Kenya?

Prof. Marcel Rutten

Introduction

Worldwide, the use of water is rising faster than the growth of the world population. In the 1900-90 periods the world population increased from 1.7 billion to 5.5 billion people while the total consumption of water went up by a factor 10 from 500 to 5,000 cubic kilometers. This explosive rise is not just due to a higher human consumption of water. The consumption of food, roughly estimated to be 1 litre of water per calorie and the improvement in water provision should be held responsible for this. In the period up to 2050 the total human population on earth is expected to grow to total some 8.5 billion. Africa, especially, will see a population boom resulting in this continent becoming the most water stressed continent. In addition, climate experts are warning that large regions of Africa will receive less rainfall in the decades ahead. However, whether people will have access to water is not just an issue of a physical or even financial nature. Other, maybe even more pressing, issues are at stake as well.

In this article, our attention is directed towards the way political, juridical and economic processes interact and contribute to an upsurge of the pressure on natural water sources and ground water reserves available in Africa. This will be done in particular by portraying the water situation in semi-arid Kajiado District in southern Kenya. Our attention will focus on a comparison of the variety of efforts proposed and implemented to solve the numerous problems in water provision in the area. In particular, we will discuss two types of water facilities: boreholes and shallow wells.

Water Problems in Kajiado District

The availability of water in Kajiado is of major importance. First, the erratic and unreliable precipitation is a major limiting factor for practising cultivation and for keeping livestock. The hills and valleys have a significant effect on the annual rainfall. Near the hills precipitation is high (800-1,000 mm), while in the low lying savannah we experience less rainfall (300-500 mm). Second, the water structures in this area are varied and include (perennial) rivers, natural wells and depressions, human made reservoirs (pans), dams (up and sub-surface), modern and traditional shallow wells, boreholes and piped water.

The Maasai pastoralists who have inhabited this area for centuries consider the lack of water (*enkare*) primarily as a problem for their herds and less as a matter of concern for human consumption. The herders and their livestock will in general settle within a radius of 5 km from a water point. During the dry season, though one has to search for new pastures. The availability of groundwater in the neighbouring areas during these dry seasons is crucial as the geographical distance and physical condition of the animals necessitates drinking every other or even third day.

During the wet season the animals will spread out, making use of the new grass and fresh surface water in the pools, riverbeds, dams etc. When the dry season returns, livestock will once again return to the neighbouring wetlands and higher terrain. It is in precisely these areas that competition with other economic activities, cultivation and wildlife parks, has increased in the last decennium.

History of Maasai Land Ownership in Kenya

The history of the Maasai is one of a constantly decreasing territory, both in quantitative and qualitative sense. In the footsteps of the Scottish geologist and explorer Joseph Thomson – the first European who managed to cross Maasai land in 1883/4 – Britain and Germany struggled over the hegemony of this area. The outcome of this battle was a subdivision of the Maasai territory. The northern Maasai were placed under the rule of the British 'East Africa Protectorate' and the southern Maasai under that of German Tanganyika. The British authorities signed two treaties with representatives of the Maasai in 1904 and 1911, respectively. As a result, the Kenyan Maasai saw a reduction of their pre-colonial territory, measuring some 60-70,000 km^2, by approximately 40%.

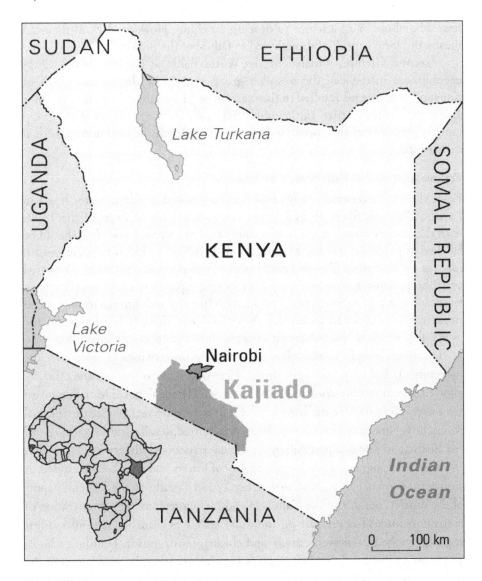

Kajiado District has been subdivided into three new districts since 2009

The losses in a qualitative sense are even more serious. The Reserve set aside for the Maasai in southern Kenya had vast areas characterised by lack of water and grass. The boundary of the Maasai Reserve is drawn in such a way that most rivers run within the territory that was in the hands of the colonialists. Protests by the Maasai, at one time supported by a British lawyer, a few colonial administrators and British investigative commissions were in vain. At best, some minimal adjustments in these boundaries were made, or the Maasai were allowed to make use of their former pastures and water sources in times of extreme droughts. After

203

World War II, the colonial authorities attempted to persuade the Maasai to agree to a system that sets a limit to the maximum number of families allowed to live near a borehole in an attempt to develop ranching. However, lack of financial means and long periods of drought led to failure of the project.

A second attempt, initiated by the World Bank, at the end of the 1960s encompasses introduced the so-called group ranches in the northern part of Kajiado District. This resulted in the creation of 51 communal ranches, which is about 75% of the district. In the mid-1980s, calls for the abolition of the group ranches grew louder and appeals were made to subdivide the land into individual ownership.

Pressure on Available Water Sources Increases

From the 1970s onwards, individual Maasai pointed at falling water levels in Kajiado District's rivers. Indeed, when comparing a list of rivers of the 1930s with the current situation, we can conclude that more rivers nowadays should be labelled as perennial and less as permanent. It is hard to know the exact reason for this development. But one could point at causes such as mining, irrigated agriculture, deforestation and the loss of storing capacity in the rivers because of river sand mining. A growth of the population (natural and through immigration) needs to be mentioned to explain these land use changes impacting on the availability of water and not simply attribute it to the effect of climate change.

The water supply is also affected indirectly by activities of outsiders, the government, industry and agriculture. Firewood and river sand are officially allowed to be transported out of Kajiado District. On the slopes of Mt Kilimanjaro trees were cut to provide the army with wood in the early years. Since the colonial times, huge amounts of sand have been transported, legally and illegally, out of the district, to be used particularly in the construction industry in Nairobi. In this way, the water conserving functionality of forests and river beds is eroded. A similar problem of logging is occurring, legally and illegally, on the other hilly spots of the district, notably in the Namanga and Ngong regions. The immigration of cultivators intensifies pressure on the water resources available. A similar threat is posed by the new flower, ostrich, and chicken farms, private boarding schools and training institutes that have emerged in the recently subdivided areas of the district. All of these activities have a high demand for water. Careful monitoring is needed to be able to determine if the above-mentioned developments will indeed lead to the much feared decrease of ground water reserves.

More clarity seems to exist towards the loss of water supplies, especially since the 1980s, in the region bordering Mt Kilimanjaro. The swampy areas in the lower parts are decreasing. In addition to the pastoralists who are confronted with a loss of dry season grazing areas, irrigating agriculturalists in this area are heavily dependent on the water of the Noolturesh river. All of them fear that the decrease in water availability will threaten their existence. Responsible for this trend is the Noolturesh water pipeline project. This project, supported by the Italians, was launched in 1992 and was immediately dubbed a 'White Elephant' because of its size, construction and the high water use. Now Kajiado Town receives some 11% of the Noolturesh water. Machakos Town, in the neighbouring district, that does have alternatives, gets 66% of the total supply. The remainder goes to the Athi River that gets enough water from neighbouring boreholes. Technical advisors have warned that the Noolturesh will not be able to meet the growing demand of these three towns. Extra sources have to be found, while the 262 km long pipeline should provide water to the pastures primarily, with only an additional supplementing function for the urban region. The local Maasai hardly profit from this mega project.

As a result, Maasai herders in search for water, destroy the pipeline. They do this partly out of frustration over the limited number of taps available and because of the high water bills. These uniform water payments, based on livestock numbers and not on the real consumption of water do not give an incentive to safe water. On the contrary, this arrangement at one time has led to the creation of irrigated cultivated fields alongside the pipeline.

Finally, another recent phenomenon is the fast reduction in the quality of water in the intensively cultivated regions of the district, especially in the Ngong region. Boreholes have been constructed in a small area resulting in brackish borehole water. Also the use of pesticides and fertilizers in the cultivated areas pollutes drinking water for both human and livestock consumption downstream.

The Ewaso Ngiro South, for example, carries pesticide residues from the large-scale wheat farms in the neighbouring Narok District. In addition, irrigated agriculture in the Rombo and Isinet areas of south east Kajiado have been blamed for polluting water downstream, affecting both man and beast. And, what will be the effect of the use of pesticides from the horticultural industry?

Complaints have been aired to the National Environmental Management Agency (NEMA) over this in the Isenya area. Also mining and small leather industry activities in Athi River pollute the streams in Kajiado District as does the waste from slaughter-houses in urban settings.

Water Facilities in Kajiado District

In 1960, Kajiado District had 38 boreholes and over 50 dams owned by the African District Council, the central government, private enterprises and individual Maasai. In the beginning of the 1960s, the Maasai were hard hit by a major drought and they lost huge numbers of livestock. This and the earlier mentioned fear that Kenyan independence would result in a possible loss (of control) of their land, made the Maasai finally agree with the creation of group ranches in the framework of the 'Kenya Livestock Development Project' (KLDP). An important argument for the Maasai to accept this World Bank funded project was the promise that substantial attention will be devoted to improving water facilities in the group ranches by way of drilling boreholes and the construction of dams, troughs and water pipelines.

TABLE 1: WATER FACILITIES IN KAJIADO DISTRICT, 1988

	Division											
	Central		Loitokitok		Ngong		Magadi		Total Group		Total Kajiado	
FACILITIES	F	NF	F	NF	F	NF	F	NF	F	NF	F	NF
Boreholes (MoLD)*	33	23	8	3	4	7	-	1	45	34	82	45
Boreholes (MoWD)	45	93	10	22	27	30	1	4	83	149	139	248
Water pans	84	-	2	-	4	-	-	-	90	-	135	-
Water dams	20	2	4	-	5	1	9	1	38	4	50	7
Wells*	10	-	8	-	-	-	-	-	18	-	29	-
Springs	13	1	17	-	3	3	-	-	33	4	44	4
Rock Catchments	4	-	2	-	17	-	20	-	106	14	170	14
Water tanks	57	14	12	-	17	-	20	-	106	14	170	14
Km water pipeline	129	18	87	-	113	-	85	-	414	18	574	18
Troughs	113	7	13	-	13	-	19	-	158	7	217	9

Source: MoLD, 1988; Mwangi, 1990

F = Functioning, NF = Not Functioning, GR = Group Ranch territory, KD = Kajiado District underestimate.*

The figures refer to the facilities in the group ranches of Kajiado District, excluding individual ranchers. The latter facilities are included in the Kajiado District total. The borehole figures of the Ministry of Livestock Development (MoLD) are far below those of the Ministry of Water Development (MoWD) (1988). Not-functioning boreholes are either not operational (144), disbanded (21), dried (12) or not traceable (71).

In the first phase of the project, from 1968 to 1974, water facilities doubled. However, in 1974, the Maasai lost access to one of their most crucial dry season grazing areas: Amboseli. This year-round green and well-watered pasture set aside for survival in times of severe drought was turned into a National Park, by the Kenyan authorities, mainly due to pressure form international conservation lobbying. As a result, the area is no longer available to the Maasai and their herds. Incidentally, a major drought began that would last until 1977. In this year, two boreholes were availed as part of compensation for the loss of water sources inside Amboseli. Unfortunately, the design of the boreholes was not that effective or sustainable and ended up being very expensive to maintain.

During the drought of 1984, the Wildlife Department was not capable of pumping water outside the Park because of budgetary restrictions. By way of compensation, the Maasai are allowed to enter the Park to fetch water for their animals. Likewise, problems in the group ranches increased in the second phase of the KLDP and many pumps and spare parts were stolen. In one of the group ranches, Erankau, members were forced to repay a loan for a few boreholes that had been drilled in an area where no ground water was found. Table 1 provides an overview of existing water facilities in the group ranches in 1988. It is important to keep in mind that a majority of these have not been constructed through KLDP, which constructed only 15 boreholes, 19 tanks and 20 troughs. Other donors also provided funds to drill boreholes and construct other water facilities in the group ranches.

As mentioned earlier, access to ground water is of crucial importance to livestock during the dry season. Groundwater is available from natural sources, shallow wells and boreholes. Natural well water is often of excellent quality but many of these sources are in hilly or even mountainous terrain. To get access to this water often means building a network of pipelines which is quite costly. Boreholes and shallow wells are often a better alternative. Most of the wells in the district can be found in the dry riverbeds and along the riverbanks. The depth of a shallow well varies from three to thirty metres. The average depth for a borehole to strike water is 80 metres, yet also at depths of 250 metres water has been found.

Development, Ownership and Operation of Boreholes

The development of boreholes in the district is foremost a business of donors, the government and the churches. In addition, wealthy individuals have been drilling their own boreholes while others have organised themselves into groups. The costs involved in the drilling and equipping of a borehole (100 m deep) are on the higher end and come to (early 1990s) some Ksh. 2 million, approx. US $40,000.

TABLE 2: YEAR OF CONSTRUCTION OF BOREHOLES 1927-93 PER DIVISION IN KAJIADO DISTRICT

Period	Ngong Division	Magadi Division	Mashuru Division	Central Division	Loitokitok Division	Total
1927-30	11	-	2	-	1	14
1931-40	-	-	1	-	-	1
1941-50	7	2	8	19	3	42
1951-60	15	-	1	13	6	35
1961-70	2	-	7	18	6	33
1971-80	19	3	10	46	12	90
1981-90	96	1	3	74	26	200
1991-93	30	-	20	3	-	53
Per year	2.7	0.1	0.8	2.6	0.8	7.0
Total	180	6	52	173	54	468
%	39%	1%	11%	37%	12%	100%

Source: Mwangi, 1993

Table 2 provides a detailed overview of borehole construction based on the borehole-register of the Ministry of Water Development. The rise in borehole numbers since the 1970s is the result of Kajiado District losing its status of 'closed district' at the end of the 1960s. Many of the new boreholes are in the hands of non-Maasai immigrants, schools and the government.

Some of the most recent immigrants are prominent national figures who seem to have bought the land for either residential or speculative reasons. Another reason for the increase in boreholes in Kajiado District is politics particularly towards the end of the 1970s. Politicians, in particular Maasai ministers, transferred large amounts of money to the Kajiado water sector, especially in Loitokitok and Mashuru divisions and other Maasai politicians would follow this example in latter years.

Unfortunately, the borehole register does not show information on the operational condition of the boreholes. To answer this question, the Arid and Semi-Arid Lands Kajiado District (ASAL) project, a Dutch funded integrated rural development project, financed an exercise to collect this data in the whole district in 1988. The major objective was to arrive at an up-to-date overview of property and condition of the boreholes. It was concluded that in the period of 1927 to 1988, a total of 387 boreholes had been drilled in the district. Out of these only 139 (36%) were still operational in 1988. Since the 1990s, the drilling of boreholes, especially by immigrants, has increased for domestic or commercial purposes. Since no district-wide survey has been conducted in recent times, the figure of less than 400 boreholes still features in current official publications and there is still no information available on their condition. Yet we estimated that, especially as a result of immigration and a boost in residential plots in the district, this number has increased possibly to over 1,400 boreholes.

Failure of Boreholes

The construction and maintenance of boreholes can at times create major problems of technical, physical as well as socio-economic character. The diesel pumps have to be supplied with diesel and oil daily which is no easy feat considering the isolated areas the boreholes are located in. Sometimes, (contaminated) cheap diesel is bought which ruins the engines.

Most of the solar panel - driven boreholes also are no longer in operation due to thievery of the panels and damage caused, especially by youngsters throwing stones. Moreover, it has become clear that the enormous demand for watering the animals cannot be delivered in sufficient quantities by solar power - run pumps and within a limited span of time. This technical limitation increases the waiting time at the borehole in such a way that insufficient time is left to graze the herds in far-away locations.

Problems with wind energy are mainly in repairing specific parts of the windmill. This is costly and takes a lot of time. Also the problem of the so-called 'overflow' is a point of concern. Large amounts of water flush away to natural depressions and this standing water is a potential health hazard as it is an ideal breeding ground for diseases. Furthermore, all types of boreholes have to deal with challenges of the relief. The geology of the district is such that cracks in the lower ground layers can cause water streams to redirect without warning. As a result, the yield of a borehole might be drastically reduced and affect the quality of water where it becomes too salty in some regions. This has also been caused by uncontrolled over-pumping of water where too many boreholes in small areas work for long hours.

In addition to these technical and physical problems, management issues have also arisen in regard to boreholes. The introduction of boreholes in the Maasai community was mainly initiated top-down. The local community was hardly involved in the choice for location, the installation, repairs and maintenance of the new technology; a situation that has been replicated to-date.

Searching for Alternatives

During the ASAL borehole survey, attention was given to alternative water facilities. The pan and the shallow well turned out to be of major importance for the rural population. These two facilities were among the most used water sources whereas the boreholes were only able to provide 2% of the total local demand for water. This situation made the Kajiado District water officer, in the late 1980s, decide to collect more information on the ways the Maasai resolve the issue of lack of water. A conservative estimate put at least three quarters of the water demand for livestock as taken care of by shallow wells, pans and dams.

Maasai pastoralists have in the past indicated a willingness to invest considerable sums of money in the development of these water sources, especially shallow wells due to the relatively low operational costs and the durability and reliance of these wells. However, they pointed out some negative aspects of the shallow wells with silting as one of the main problems, especially during the rains. Sand storms also fill the wells with sand and this can translate into a lot of time and money spent on repairs. Other problems include the potential collapse of the sand walls and the pollution of the water as the wells are not covered. This implies that all kinds of vermin, (livestock) chemicals, urine and faeces of the animals accumulate in the wells especially during the rains.

In this light, it was decided that data would be collected with the main aim of finding ways to improve the design, operation and management of the wells. In 1993, a survey was conducted for each single shallow well in the district to ascertain its location, construction, the quality of water and the manner in which the water is drawn. A pilot study was also carried out to improve a limited number of traditional shallow wells by deepening them, putting filters, lining and strengthening the walls, covering the well and finally placing and testing a number of hand and mobile diesel pumps on top.

Shallow Wells: The Maasai Solution

Usually, a number of Maasai herders meet with their herds near a well early in the morning. The water is lifted in buckets using a "human ladder": 6 or 7 people standing above each other in manufactured inroads in the wall of the shallow well. The buckets are then emptied into a nearby trough and the animals are led, in small groups, to drink from the trough. The number of families using a single well depends on the season and its capacity, but it generally varies from 2 to 20.

The Maasai, like most nomadic pastoralists, make use of the dry riverbed to find water during the dry season by scooping the sand. In the past water, was lifted from these temporary wells using animal skins. A huge disadvantage of this type of well (o-sinyai) is that after the commencement of the rainy season, the sand covered the well again, or at the worst, made future use of the well impossible due to the strong currents. In addition to the o-sinyai type there are also wells dug in the riverbank either in sand or in stone. Almost all of these wells (o-lumbua) have been made by the Mbulu ethnic group. Certain members of this group, originating in Tanzania between Lake Eyasi and Lake Manyara, are moving all over Maasailand constructing these wells in return for a payment.

An important aspect of these wells is finding the right location to sink a well which requires expertise. The presence of certain tree species: Oltepesi (Acacia Seyal) and Olerai (Acacia Tortilis) is a sign that water at great depths will not be found. The time and costs for construction of a well depends on the type of the well, the depth needed to reach the water, the desired volume of water and the solidness of the rock or soil. Digging is done through the use of *jembes* and picks and the trough is made out of mud and cement. Upon completion, the well is fenced with thorny branches to keep out wild animals. An *o-sinyai* well (3-5 metres deep) is ready after about one month. The *o-lumbua* may take 3 months to a full year (10-30 metres deep). In case funds are not enough, the time needed may take much longer. All in all, the total construction and maintenance costs of a (traditional) *o-lumbua* well are few compared to a borehole's.

Marriage of Traditional and Modern Knowledge

The ASAL programme well count totalled to 1,505 wells with the oldest well, dating from 1920, still in operation. The ASAL programme took the lead in the implementation of the practical phase of this Shallow Well project. The project assists, on a cost-sharing basis, the financing of the improvement of the wells. The improvements done includes placing a filter, enforcing and lining the walls, covering the well and placing of a hand pump. These pumps have proven to be able to produce much more water than the Maasai human ladder. As had been mentioned earlier, District Commissioner Wainwright had rightly expressed in the 1940s that the Maasai would welcome the hand pump with great enthusiasm.

The ASAL programme has trained the well users in the operation and maintenance of the pumps and where to get supplies for a pump ensuring that the users know where to go if the pump breaks down. This makes them fully responsible for the maintenance and repairs of these modernised wells in the same way they took care of their traditional wells in the past. It was decided that some wells would have two pumps: one hand pump to collect water for domestic purposes and another mobile diesel driven pump to draw water for the animals. This separation within the water drawing system allows for a higher quality of domestic water because the water is no longer taken from troughs. Moreover, the women no longer have to wait their turn, after all the animals have had their fill, before fetching water.

A second change in the use of shallow wells was initiated by non-Maasai immigrants who have settled in their newly acquired plots of lands. Due to the need for water, they have dug wells next to their homes and they have struck water at varying depths of 5 to 30 metres depending on the location of their land. They showed the Maasai that water was not restricted to the river and its nearby surroundings. It is possible too that the drilling of boreholes by the authorities had the Maasai reconsidering these alternatives and they also began to dig wells in their ranches. Many Maasai have combined efforts, through grass roots organisations, to finance and provide labour in digging or improving shallow wells.

By the late 1990s, the number of shallow wells in Kajiado District had risen to over 3,000. The water drawn from these sources opened a new set of opportunities. Some people placed water tanks next to the well to store water and cement, aluminium and lately large numbers of black plastic tanks or combinations of these are sprouting all over the sub-divided group ranches of Kajiado District. These tanks serve as storage for the dry season. Water is transported, through a piped system, to irrigate agricultural plots, to water troughs for livestock and to water taps for domestic use. This has really revolutionised the economic situation for a number of households who grow tomatoes, onions, vegetables, with a few Maasai venturing into the cultivation of bananas, oranges, mangoes, paw paws and even grapes.

Often, non-Maasai are hired to work in these *shambas* (farms) but more Maasai men and women are beginning to participate actively. For the latter, this development in the water sector has significantly reduced the time and effort needed to fetch water which can now be spent on other constructive activities. Food security seems to have significantly improved because of diversification of economic activities. In addition to new opportunities in cultivation, another promising area is now open for long term opportunities: the growing of special kinds of high quality grasses. As far back as the 1930s, the colonial authorities had already experimented in this field. As a result of the sub-division of the group ranches and due to the availability of water, it is now possible to grow grasses such as *Napier* and *Bana* grass. The first results seem to indicate that a combination of 'run-off water harvesting' and some extra irrigation during the height of the dry season allow these grasses to survive. As an extra fodder, this could lead to five times higher milk yields. With a rapidly decreasing availability of land for rearing livestock, this might be the best solution to pastoralism.

Shortly after the turn of the century, when the ASAL programme wrapped up its activities in Kajiado district, it was concluded that the collaboration between ASAL officials and well owners had resulted in an improved well design that promised to be sustainable from a financial, technical as well as management point of view. This new approach has overcome many past failings in the water sector, such as improper or misplaced technological improvements, unclear ownership and poor management. Throughout the 1990s, all the water points operated smoothly, the wells were effectively managed and their numbers increased providing water for livestock, cultivation and human consumption. Even during the extreme drought of 2000, the wells still had water. Food security, in particular, was boosted by the rapid spread of small-scale cultivation. However, from 2003 onwards, reports trickled in that some newly developed improved wells had started to dry up and in 2007, the decline in water levels continued. The former ASAL water engineer responsible for this innovation decided, in collaboration with the author of this article, to revisit the improved wells to monitor their current state and determine the cause of this trend.

Preliminary Findings from the 2007/8 Revisit

The revisit showed that the development of improved shallow wells, which began in the 1990s, had continued. Two aspects, however, were seen to be wanting.

First, the development of new wells was more problematic after the ASAL programme stopped operations. Financial support and technical advice were not easily available. There was a threat posed to already existing improved wells by a number of new developments that had occurred in the area. New large scale settlements and commercial agricultural activities had resulted in the sinking of many deep boreholes that were draining the aquifers and reducing the quality of the water. While these new intrusions differ in type and magnitude, they are all working to the detriment of groundwater sustenance. The worst effect to date has been the drying up of a large number of the (improved) shallow wells near these new activities, forcing the owners – the majority of whom are relatively poor – to buy water from their new commercial neighbours.

In total, there are 105 deep boreholes supporting different endeavours in the Olkinos and Embolioi study area. Since communal land was sub-divided in the mid-1980s, it has become a commodity that allows new landowners to settle there. Horticulture and poultry farming are becoming important, with these two enterprises accounting for 32 deep boreholes that, due to over abstraction, have led to the drying up of 29 shallow wells by mid-2008. Most of the boreholes are within the prohibited distance of 800 m from any other source of groundwater. Lack of control and monitoring by the authorities is responsible for this breach of the Water Act. The area's flower farms are also being blamed for polluting water courses and for livestock disease and deaths. Environmental policies are ineffective in prohibiting the destructive exploitation of natural resources. The recent introduction of eucalyptus tree farming (for electricity poles, construction and firewood), motivated by its potential as a good earner and its fast expansion, is likely to reduce water tables further. The Olkinos/Embolioi area accounts for 14 eucalyptus farms and these have contributed to the drying up of 3 shallow wells in a radius of 0.1 and 1 km within two years of the trees being planted.

Another threat comes from sand and vegetation harvesting in the seasonal rivers of Kajiado. Some riverbeds are now bare rock, with the shallow wells in and along these seasonal waterways being lost as a result of unabated sand harvesting. The sand, which holds and protects the water for the shallow wells and aquifer recharge, is being exported outside the district to the construction industry in Nairobi and beyond. The river banks have become bare of trees due to charcoal burning and higher levels of water evaporation. The threat is not only to people and their economic activities but also to wildlife and the ecosystem in general.

Finally, with the growth of Nairobi, more people are moving into Kajiado District. Kitengela town and its surroundings have grown tenfold in the last ten years due to immigration. As more settlement takes place, wells are being drilled to meet the rising demand for water. These new challenges are undermining the improved water wells and present an inconvenient scenario for all stakeholders. Some may question whether the drying of wells is caused by climate change. Orindi & Murray blame governments in East Africa for not preparing for climate change which they claim is already widespread in the region with dry lands becoming drier. However, when checking rainfall figures for the semi-arid Olkinos/Embolioi region, we concluded that since 2000, average rainfall is up from 603 mm for the long mean in the 1962-2007 period to 622 mm per year. Out of the 9 wet years in the 1962-2007 period, 5 occurred in the last 11 years, with number of rainfall days remaining constant (approximately 50 per year).

Out of the 5 dry years, 3 occurred in the 1970s, and one only (2000) in the last 11 years. Orindi & Murray further claim that the best way to mitigate climate change is not in the technical solutions proposed by national governments. 'Real, long-term solutions can be found in existing livelihood strategies used by communities. Recognition of traditional land rights and systems of management in the formal land laws (for example in Tanzania) is therefore a significant step towards reducing vulnerability of communities.' Experiences in Kenya, however, as shown above, raise questions over their claim since formalisation leads to commoditization triggering the introduction of less sustainable alternative land uses.

In addition, water use is officially regulated in the Water Act. However, this presupposes that the Ministry's field officers have the resources and know-how to assess water use and local demands and enforce regulations. Resource constraints within the public sector limit the effectiveness of such mechanisms. The Act does, however, provide limited regulation of extraction but is unable to address broader environmental issues that could impinge upon issues of water quality or overexploitation.

Conclusion

In this article, we stressed the importance of water for the Maasai economy. Inhabiting the semi-arid plains of southern Kenya, the Maasai pastoralists are eager to develop sustainable water sources. These sources and the accompanying machineries have over the years become less easy to obtain. This is due to an autonomous growth in demand for water from a growing population and larger cattle herds and also because of intense interference through political and juridical processes, from past to previous times.

Research has shown that shallow wells, the local way of water collection, have always played a significant role in the management of natural resources by the Maasai but that top-down implementation of modern technology, both by the government and foreign donors, had the planners forgetting about it until the early 1990s. For too long, the donor community only valued modern, large-scale and expensive techniques. Local, small-scale and cheap alternatives were considered inferior and less reliable. It is the longstanding experience and interest from the side of Kenyan water specialists, through funding by a Dutch donor, that re-awakened the interest in and knowledge of shallow wells. Moreover, due to a parallel change in land rights in the Maasai area of Kajiado District i.e., the sub-division of group ranches into individual plots, the construction of shallow wells rose significantly.

Tailor-made modern knowledge was applied to the traditional shallow wells to produce improved versions of the wells. This modern knowledge came from both trained technicians and non-Maasai immigrants. Likewise, the traditional knowledge is a mixture of wisdom from Maasai and Mbulu water specialists. As the Maasai saying goes: *Metolu lung' elukunya engeno* - one single head does not have all wisdom, more heads are better. This collaboration resulted in an improved well in such a way that it is now beneficial and sustainable from a financial, technical and management point of view. The passive consumption of western techniques was stopped and exchanged for one that builds upon the resources, means and ideas of the local people. From a role of supplying and donating resources, the donor has moved into a position of translator and facilitator.

However, formalisation of land tenure arrangements impacted on the water use in the area because newcomers who moved into commercial (export) agriculture, notably horticulture and tree plantations decided to drill boreholes indiscriminately. Scooping of sand from rivers and destruction of riverside vegetation also negatively affects the water holding and recharging capacities of the area. The Maasai and other small-scale immigrants in the Olkinos and Embolioi areas who had progressed in food security and gained economic wealth after shallow wells were improved are now witnessing powerful neighbours lowering water tables and subsequently frustrating the progress the former had gained over the years. Slightly increased precipitation over the last decade and questioning claims of drying of the semi-arid zones of this part of Eastern Africa cannot override these assaults on the water table.

Verbal and written complaints from owners of shallow wells affected by the drilling of deep boreholes have so far not been heeded. It cannot be ruled out that disillusioned individuals and communities may in the future take the law into their own hands to 'solve' the problem. Through the Water Act, the Ministry of Water and Irrigation is entrusted with regulating water access and use but has failed to do so. Besides maintaining and strengthening existing laws, commercial farmers should be held responsible for the unsustainable groundwater abstraction practices that are threatening the environment. These short-term exploitative practices may ultimately destroy the employment opportunities and economic wealth that has so far been created. Unless the outside threats on quantity and quality of water sources in Kajiado District are kept at a minimum, the marriage between traditional Maasai and modern knowledge will not last long.

Notes

1. Falkenmark, M. & J. Lundqvist. 1994. "Development in the Face of Water Shortage". SIDA-Infrastructure Division, Stockholm.

2. Norton Griffiths, M. 1977. "Aspects of Climate in Kajiado District". UNDP/FAO Kenya Wildlife Management Project. Project Working Document 13, Nairobi.

3. The British colonial power tried to interest Europeans to settle in Kenya. Part of the Maasai territory was offered to Jews in Eastern Europe. The Jewish World Congress, August 1903, however, decided not to accept the offer, but instead to strive for an autonomous state in Palestine. Still the Maasai had to make room for other newcomers, mostly South African Boers.

4. In front of the Carter Land Commission the Maasai once more repeat their grievances about the loss of their dry season grazing areas and water sources. They had every single reason to be bitter because the northern boundary of the Maasai Reserve 'was so drawn as to exclude the most valuable water supplies, which are included in the land alienated to Europeans'. James, L., "The Kenya Masai: A Nomadic People under Modern Administration," *Africa, 12*, 1939. pp. 49, 73.

5. Rutten, M.M.E.M. 1992. *Selling Wealth to Buy Poverty: The Process of Individualization of Landownership Among the Maasai Pastoralists of Kajiado District, Kenya, 1890-1990*. Saarbrücken: Verlag Breitenbach Publishers.

6. KDAR 1929; KNA/DC/KAJ.3/1 HOR, Kajiado District Annual Report, Nairobi: Kenya National Archives, January 1946. p. 17.

7. Daily Nation, December 10, 1975. There were reports of illegal logging going on in the eastern part of the County Council of Loitokitok forest. Immigrants fell trees for construction, sell as firewood or burn charcoal. It resulted in the reduction of the of water in the Kuku plains. The felling of trees for charcoal was prohibited, but in April 1976, it was allowed again within the framework of the 'plant by cultivation' project.

8. Soini, E. 2005. "Land use Change Patterns and Livelihood Dynamics on the Slopes of Mt. Kilimanjaro, Tanzania" in *Agricultural Systems, 85*, pp. 306-23.

9. Campbell, D., Lusch, D. Smucker, T. & Wamgui, E. 2003. "Root Causes of Land Use Change in the Loitokitok Area, Kajiado District, Kenya", LUCID Working Paper 19, p. 20.

10. In April 1963, the Kajiado Agricultural Committee formulated a memorandum for a WHO/FAO-Mission concerning the development of water sources in the Maasai area. It mentioned the necessity to chose a long term vision with respect to the provision of water, improved coordination of relevant water plans between all of the seven government bodies involved in water development in the district. A more sedentarised way of life of the Maasai pastoralists is predicted which would mean a change in the use of the land. It was concluded that the dams have turned to be non-sustainable and for that reason improvement of water extraction from the river is promoted through the improvement of wells in the dry river bed, the development of shallow wells and water harvesting roof catchments in the higher areas, the extension of pipelines and boreholes.

11. Ministry of Livestock Development Kajiado District *"Kajiado Livestock Census: Integrated Livestock Census and Infrastructure Survey Report,"* Department of Livestock Production Kajiado District/Arid and Semi-Arid Lands Programme Kajiado, 1988.

12. Mwangi, M.N. 1990. "Water Development in Kajiado District: Some Personal Remarks," in Klinken, M.K. van & J. ole Seitah (eds.). *The Future of Maasai Pastoralists in Kajiado District (Kenya)*, Proceedings of a Conference held in Brackenhurst Baptist International Conference Centre Limuru, Kenya, 28-31 May, 1989. ASAL Programme, Kajiado, pp. 114-19.

13. Ministry of Water Development, *"Kajiado Borehole Survey"*, funded by Arid and Semi-Arid Lands Programme, Kajiado District, 1988.

14. Njoka, T.J. (1979, Ph.D. thesis). "Ecological and Socio-cultural Trends of Kaputiei Group Ranches in Kenya". Berkeley: University of California. p. 181.

15. Moss, C. 1989. *Onder Olifanten. Veertien jaar met een Afrikaanse olifanten Familie*. Amsterdam: De Boekerij bv. p. 230.

16. Livingstone, I. 1986. *Rural Development, Employment and Incomes in Kenya*. Aldershot: Gower. p. 271.

17. Peron, X. 1984. "Water Policy in Maasai Country in Kenya: Development Without Participation," presented at the symposium, *'Agricultural Development and Peasant Participation: Water Policy'. 'Bulletin de Liason', CREDU, Newsletter, issue 12,* pp. 41-71, 61.

18. Dietz, A.J., A.P. Owiti, J. Brandt & J.O. Atinga. 1986. "Report of the Identification Mission for an Arid and Semi Arid Lands (ASAL) Programme in Kajiado District". Prepared for SNV/ Netherlands Development Organization, Nairobi. p. 12.

19.

20. Ecosystems Ltd. 1982. "Amboseli/Lower Rift Regional Study, Final Report". Report prepared for the Wildlife Planning Unit, Ministry of Tourism and Wildlife, Republic of Kenya. p. 13.

21. Mwangi, M.N. (1993, MPhil thesis). Research Report on Shallow Wells in Kajiado District, Kenya. Loughborough University, UK.

22. Klinken, M.K. 1993. "Maasai Pastoralists in Kajiado, Kenya: Taking the Future in their own hands?" presented at the joint IAGIA-CDR Conference on *'The Question of Indigenous Peoples in Africa',* Copenhagen, Denmark, 1-3 June, p. 9.

23. The ASAL Kajiado programme had an annual budget of 1 million guilders. About 10 to 20% hereof was spend on water projects. Attention is mainly directed towards borehole rehabilitation, rural water supplies, micro water supply (shallow wells, tanks, roof catchments), sand and subsurface dams (artificial aquifers in rivers) and training. In 1992, Kshs. 1.5 million was spend on borehole rehabilitation and rural water supplies and Kshs. 1.9 million to micro water supply, sand and subsurface dams and training. The principle of cost sharing was introduced in 1993, where 50% of the costs are expected to be paid by the community.

24. Mwangi, 1990. There are 45 boreholes in the group ranches of which only 23 are operational. Individual ranchers posses 155 boreholes (56 functional). Institutions such as livestock training centres (59bh/22f), Ministry of Water Development (54bh/23f), other departments (27bh/2f) and the Kajiado County Council (37bh/13f) are co-owner of the other boreholes. Often, the owner of the non-operational boreholes is usually unknown. This borehole water is primarily for human consumption in villages, schools and health centres. p. 115.

25. In 2009, the National (Nairobi) borehole registers for Kajiado and those at the district level had conflicting figures. The likely number of boreholes, in the "old" Kajiado District, stands at 1,496. This is a growth of over 200% since 1993.

26. Specific information on this research, courtesy: Dr Moses Mwangi.

27. The 1946 Kajiado District authorities' report on the Kamba and Mbulu well-diggers employed by the Maasai. To prevent these people from settling permanently in the "closed" Maasai District, it was announced that these labourers should request for passes for the duration of their stay (KNA/ DC/KAJ.3/1 HOR September 1946:4).

28. The Maasai in the past, during their move southward, regularly raided the Mbulu for cattle. The weaker Mbulu developed a method to hide from the Maasai; they dug underground hide-outs for them and their animals, mostly in hillsides. While digging these hide-outs, the Mbulu often struck groundwater. Subsequently, they specialised in digging shallow wells in the riverbanks. The Maasai decided to put down their weapons in return for support from the Mbulu in their search for water. Nowadays, Maasai pastoralists still prefer to leave this intense manual labour to neighbouring groups like the Mbulu, Kamba, Luo and Kikuyu. However, in recent times, some Maasai have taken up well-digging mainly due to poverty; whereas manual labour in the Maasai culture is considered to be for those belonging to a low social status.

29. Wagura & Kanyanjua. 1992. The 1991/92 Dutch Survey, funded by Water Resources Assessment and Planning Project (WRAP), demonstrated the neglect of shallow wells. This survey accounted for only 287 shallow wells in Kajiado. Likewise, the totals presented for other water sources, though to a lesser extent, underestimated the facts on the ground. Dutch development funds in Kajiado turned out to be better used in long-term integrated project support more than in sectoral programme assistance projects.

30. Kenya, Republic of. 1991. "Water Resources Assessment Study in Kajiado District. Inventory Report," Nairobi: Ministry of Water Development/WRAP.

31. Wagura, J.M. & Kanyanjua, J.M. 1992. "Rain Water Harvesting in Kajiado District as a Source of Water Supply," presented at the 3rd National Conference on Rainwater Harvesting Systems, Nairobi.

32. KNA/DC/KAJ.3/1-HOR, "Kenya National Archives Handing Over Report Wainwright to Wilkinson, Kajiado District, January 1946, p. 12.

33. Orindi, V. & Murray, L. 2005. "Climate Change in East Africa: A strategic Approach," *IIED Gatekeeper series 117*, p. 4.

34. Orindi & Murray. 2005. p. 15.

The Large-Scale Irrigation Potential
of the Lower Rufiji Floodplain: Reality or Persistent Myth?

Olivier Hamerlynck, Stéphanie Duvail, Heather Hoag, Pius Yanda & Jean-Luc Paul

Introduction

The Rufiji River Basin is located entirely in Tanzania and drains some 180,000 km², about a fifth of mainland Tanzania. The river has a strong seasonal flow pattern, with a flood peak around April. With an annual mean flow of 800 cubic meters per second, it is one of Africa's largest rivers after the Congo, Zambezi, Niger, Nile, and Volta. The annual rainfall is highly variable, depending on altitude and distance from the coast and generally displays two peaks with short rains in October-November and long rains from March to May. Three main rivers: the Kilombero, the Great Ruaha and the Luwegu, join up in the Selous Game Reserve to form the Rufiji which tumbles down to its wide Lower Floodplain through the 100 m deep Stiegler's Gorge before branching out in a wide delta covered by the largest stand of mangrove in East Africa (Figure 1). Administratively, the Lower Floodplain is located entirely in the Rufiji District which has some 200,000 inhabitants.

Figure 1: The Rufiji River Basin

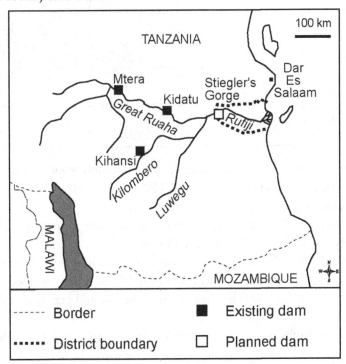

Description of Agriculture in the Rufiji District

The agricultural potential of the Rufiji District cannot be confined to the floodplain but has to be framed in a wider land-use context. Traditionally livelihoods strategies in Rufiji show a strong integration of activities between the floodplain farming system, the surrounding wooded and forested terraces and the floodplain associated lakes.[1]

In Rufiji District the agricultural potential is mainly determined by soil quality and water supply (rainfall, floods and gr oundwater). On the basis of soil type and natural vegetation cover, which is the result of the combination of soil type and the availability of water, there are four main zones in Rufiji District outside of the Selous Game Reserve (part of which covers 25% of the District but will not be considered here):

Figure 2: The Lower Rufiji Valley

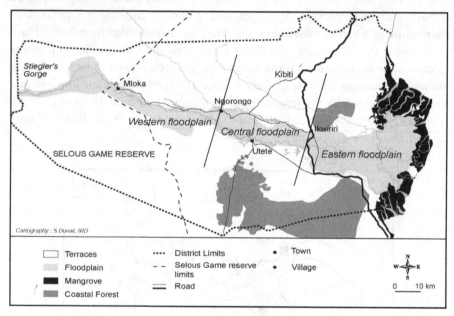

- 425,000 ha of coarse sands, covered with miombo woodland and coastal forest mosaics, lying between 50 m and 150 m Above Mean Sea Level (ASL), ('the terraces')
- 270,000 ha of finer sands (predominantly red soils), covered with coastal forest, between 150 m and 700 m ASL ('the hills')
- 170,000 ha of rich fertile alluvium interspersed with sands, covered by tall grass and occasional trees lying between 5 m and 50 m ASL ('the floodplain')
- 130,000 ha of fine salty clay, covered by mangrove and grasslands below 5 m ASL ('the delta')

The Lower Rufiji floodplain starts at Mloka where the river leaves the Selous Game Reserve. At Mloka, the floodplain has a width of some 12 km and is situated at an altitude of about 50 m ASL. From there, it gradually slopes down to the Ocean, about 120 km to the east, its width varying from 7 km (Utete) to 30 km (edge of the delta).

Floods that cover agriculturally significant parts of the floodplain occur, on average, in 4 years out of 10.[2] These floods, which transport enormous quantities of fine elements, provide the natural fertiliser to the floodplain fields (*shamba* in Swahili). The coarser sands are deposited in areas with relatively high current velocities, in the various riverbeds and on the banks when the river overflows them. When the sediment-laden floodwaters enter the plains, the current velocities drop sharply and the fine elements are deposited, as a thick layer of fertile silt. Local farmers estimate that yields are halved if a *shamba* is not flooded for 3 consecutive years.[3]

The high dynamics of the river, which continuously erodes and deposits different materials, meanders or opens new braided channels, has resulted in a patchy distribution of soils.[4] This is most pronounced in the western and central parts of the floodplain. Therefore, in any single year only a relatively small proportion of the floodplain is cultivated, at present some 15,000 ha. Fallows represent about twice as much surface area which means local farmers consider that about 45,000 ha are 'worth cultivating' in a highly opportunistic and flexible system. For an individual plot, the series of years it is cultivated continuously can vary between 5 and over 30 years, depending on the flooding frequency. In the absence of floods, yields will decline strongly after 3 years but any flood that deposits silt will extend the number of years it can be cultivated. In this manner some of the low lying plots at Ndundu have been cultivated annually for more than 30 years without seemingly any loss of soil fertility. In other areas, as shown by the ground-breaking of the village maps[5], major shifts in cultivated areas occur when strong floods open up new water courses or deposit sands or clays. With these fallows the total cultivated area adds up to about 45,000 ha.

Floodplain Zoning

There are three main sub zones in the floodplain (Figure 2):
• The western floodplain between the Selous border at Mloka and Ngorongo. In this area the floodplain is essentially situated south of the river.
• The central floodplain between Ngorongo to just upstream of Ndundu bridge. Here the floodplain is predominantly situated north of the river.
• The eastern floodplain between Ndundu bridge and the delta (saline soils). There is floodplain on both sides of the river but the frequently flooded and more easily accessible lands are essentially north of the river.

All three floodplain zones have sandy levees along old branches and braided channels that are the favoured areas for habitation and home gardens with a variety of crops (pigeon pea, cow peas, okra, pumpkin, tomato, chilli, sesame, etc.), including tree crops such as mango, banana and pawpaw. Such levees become progressively scarce towards the East as the river has, by then, deposited most of the coarse sediments and meanders rather than braids. It is important to note that these levees limit the return flow of floodwaters to the river and that, even relatively short, flood peaks result in prolonged stagnation of water on the floodplain.

The Western Floodplain

The advantages of the western floodplain are the relatively coarse texture of the fertile soils, which makes them relatively easy to work and suitable for a wide variety of crops, and its good natural drainage. Disadvantages include the very patchy nature of the soil associations and the uneven topography with levees interspersed with depressions, separated by sandy ridges. Individual farmers make excellent use of this very patchy nature but there is virtually no potential for large-scale endeavours. Rainfall is relatively low (< 700 mm) and highly variable. Most of the floodplain is quite high above the river and therefore only flooded at relatively high discharge, probably over 3,500 cubic metres. The fact that the registered villages are north of river and that the fields are south of the river sets considerable constraints on the development of harmonious communities with effective access to social services and transport.

The proximity of the *shambas* to the Selous border makes the area highly prone to crop damage and interference with agricultural activities by problem animals. Between August 2002 and April 2004, at least 35 people were killed by lions mainly on the western floodplain[6]. Fields were largely abandoned and standing crops were consumed by elephants, warthogs, etc.

The Central Floodplain

In the central floodplain, depressions with heavier soils are becoming a more dominant feature and the soil units are also becoming larger. Rainfall is higher (the 77-year average at Utete is 867 mm) and also somewhat more reliable (Table 1).

TABLE 1. ANNUAL RAINFALL AT UTETE BETWEEN 1923 AND 2002

	Annual rainfall	Number of years	Occurrence
Adequate	N > 800	46	6 years in 10
Dry year	600 < N < 800	24	3 years in 10
Drought	N < 600	7	1 year in 11
Total		77	

Source: Author's Compilation

The agricultural year was defined as September 1 to August 31 of the subsequent years for 1922 to 1980. For the years 1987 to 2002, August 1 to July 31 of the subsequent year was used. The average covering the 77 years over both periods is 867 mm. Data was collected from Rufiji District sources and BRALUP studies. Years with incomplete or unreliable data were removed.

The agricultural year was defined as September 1 to August 31 of the subsequent years for 1922 to 1980. For the years 1987 to 2002 August 1 to July 31 of the subsequent year was used. The average covering the 77 years over both periods is 867 mm. Data was collected from Rufiji District sources and BRALUP studies. Years with incomplete or unreliable data were removed.

Still, inter-annual variability is high with minimum rainfall recorded at only 484 mm (1945-1946) and maximum recorded 1,321 mm (1963-1964). Social services exist on both sides of the river and, with most of the floodplain north of the river, there are less incidents with dangerous animals, even though crop losses to problem animals (warthogs, wild pigs, baboons, etc.) are substantial.

An attempt at pumped irrigation has been made at Mbunju Mvuleni through an Iranian funded project, the Segeni scheme. The water supply comes from the outflow of the Rambo river and Lake Ruwe. Unfortunately, large flood peaks, originating from the main river, back up into this branch and have and will continue to be a substantial risk factor for any permanent hydraulic infrastructure. The scheme has therefore failed. Since 2006 a large-scale cotton farm was established at Mkongo, which is excluding access by local farmers in spite of so far not having actually planted, presumably because of low prices in world markets.

The Eastern Floodplain

The eastern floodplain comprises of the easily flooded so-called 'Ikwiriri' block on the northern floodplain. Here the heavy clay soils predominate and form near-continuous units some of which are mechanically ploughed. It is a prime area for rice growing and for *Mlao* cultivation in years of heavy or prolonged flooding. It also receives considerable rainfall (presumably over 900 mm) though at present no long-term data are available. A second major farming area on the northern floodplain is situated south of Muyuyu village.

The eastern floodplain south of the river is somewhat higher and more complex, with sandy ridges caused by the strong meandering tendency, the inflows from the Matumbi hills and the seasonal riverbeds connecting to the southern delta (Mohoro river). It is cultivated in patches. There is expansion of farms on the seasonally flooded grasslands along the new road to the bridge but the pattern is that of shifting cultivation and considerable forest clearing. Its sustainability remains to be proven. Towards the east, the floodplain grades into the inner delta. Soils become progressively heavier and more saline and cultivation is almost restricted to the levies.

The Traditional Floodplain Farming System

As from the earliest historical records, the Lower Valley of the Rufiji has been referred to as a granary for Zanzibar,[7] mainly exporting large quantities of rice and also maize. The traditional farming system is well adapted to the local conditions and rather sophisticated, though this has been recognised only by a minority of authors.[8] The floods are clearly perceived as a blessing by the local communities,[9] e.g. in the words of Mrs. Habibi Omari, one of the respected elders of Utete.[10] she '…preferred the famine caused by floods to the famine caused by drought. There is more suffering in a drought year because, after a big flood, the recession agriculture (*Mlao*) harvests are good and the fishing is good'. She also made reference to the opportunities offered by the combinations possible with two rainy seasons and the floods: '…the Rufiji floodplain has so many seasons', and finally '…the people of Rufiji have adapted to the floods'.

It is indeed admirable how the Rufiji floodplain farmers have come to grips with the interplay between short rains crops, long rains crops, floods and recession agriculture and the subtle use of the topographical variability, which defines the nature of the soils and their flooding frequency[11]. To analyse food security, under different combinations of rainfall and floods, the risk of crop failure in each of the seasons has to be examined.

Daily rainfall data are only available for a 23-year time series at Utete (1979-1980 to 2002-2003). Until more data become available from the other, recently rehabilitated, stations in Rufiji District, we have to assume that Utete is representative of the pattern over a wider area. The agriculturally relevant rainfall can be split up in the short rains 15 October to 31 January (108 days) and the long rains between 1 February and 15 May (103 days). Agriculturally adequate floods have been defined as stage board readings over 3.84m between 1926 and 1956 and peak flows of over 3000 cubic metres between 1957 and 1984[12]. 'Excessive' floods, defined as causing extensive flood damage to long rains crops, occur in about 1 year out of 7 (Table 2).

Floods	None	Adequate	Excessive
	0.60	0.25	0.15
Short Rains		Long Rains	
Inadequate	0.65	Inadequate	0.35
Adequate	0.35	Adequate	0.65

Source: Author's Compilation

These are the probabilities of the different flood and climatic events, which affect floodplain farming, occurring in any single agricultural year. For example, a 0.60 probability of no flooding means that this will occur in 6 years out of 10, adequate floods have a probability of 0.25 which means that they will occur on average in 1 year out of 4, etc. By multiplying the different probabilities, the likelihood of a combination can be calculated if the events themselves are not correlated. For example, no flood and inadequate short rains will occur in 0.60 x 0.65 = 0.39, thus in about 4 years out of 10.

Probabilities of the different flood and climatic events, which affect floodplain farming, occurring in any single agricultural year. For example a 0.60 probability of no flooding means that this will occur in 6 years out of 10, adequate floods have a probability of 0.25 which means that they will occur on average in 1 year out of 4, etc. By multiplying the different probabilities, the likelihood of a combination can be calculated if the events in themselves are not correlated. For example no flood and inadequate short rains will occur in 0.60 x 0.65 = 0.39, thus in about 4 years out of 10.

The importance of flood height is related to the topographic distribution of the soils and their aptitude for farming. The preferred floodplain farmland is on the *Mbawila* soils of the levees, consisting of loam and fine sand, relatively easy to work, well - drained and suitable for a variety of crops but in general dominated by maize. The somewhat lower-lying depressions with *Mfinyanzi* (dark heavy clay) are suitable for rice. By cultivating a number of small plots (about 0.4 ha each) with different soil types and at different topographical levels, often at dispersed locations, each household harvests from about 1.5 ha of cleared floodplain land each year with double crops in years of good rainfall and/or adequate floods. By making judicious use of their knowledge of the land they plant according to a risk - avoidance strategy, inter-cropping maize with rice on the slopes of the depressions and using different varieties planted at different times (broadly categorised as early and late). Thus, under a wide range of rainfall and flood behaviour, they are likely to achieve sustenance and in, favourable years, produce considerable surplus. Under such favourable circumstances the staple food crops are the main cash crops in Rufiji.

Land preparation, surveillance and harvesting demand considerable effort. Each household cultivates on average between 1 and 2 ha and the typical inter-cropping of rice and maize (about 60% of the rice surface area is inter-cropped) requires 367 person days per ha. Clearing and hoeing are the most demanding tasks in this. Keeping away wild animals from the cultivated plots is also very labour - intensive. The permanent presence of at least one "scarer" is required throughout the crop cycle for maize and at the early and late stages for rice. This need explains the construction of small huts on stilts called '*dungu*' at each *shamba*.

Agricultural Calendar in the Rufiji Floodplain			
Crop	Season	Planting	Harvesting
Rice		January-February	June-July
Maize	Short rains	November-December	February-March
	Long rains	May-June	August-September
Pulses, pumpkins and other flood recession crops (Mlao)		May-June	September-November
Cotton		May-June	October-November

Source: Author's Compilation

The distinction between the 'food crops' rice, maize and Mlao crops and the 'cash crop' cotton is a colonial distortion as an important proportion of the cereal production was always sold for export (in good years).

In the agricultural calendar of the main crops in the Lower Rufiji floodplain, the distinction between the 'food crops' rice, maize and *Mlao* crops and the 'cash crop' cotton is a colonial distortion as an important proportion of the cereal production was always sold for export (in good years). As can be seen from Table 3, maize is sown in November-December to benefit from the early rains and harvested in February-March, just prior to the onset of the main floods. Yields are about 700 to 1700 kg per ha. Rice is planted in January-February and harvested in June-July. Yields are about 500 to 1500 kg per ha. For both staple crops farmers in general plant different early and late maturing varieties.

In years of moderate flooding in April, the rice crop can develop even if rainfall is deficient. In years of 'excessive' floods (1 year out of 7), the rice crop is lost. However this process is necessary for the regeneration of floodplain fertility. The groundwater recharge and the deposit of fertile silt by such major floods allows for the bumper harvests of maize, planted at flood recession in May and harvested in September, the famed *Mlao* cultivation. The *Mlao* also produces other crops such as pumpkin, cow peas, etc. In the past cotton was also an important mlao crop and, at times, the main cash crop of the region.

Though virtually all inhabitants using the floodplain identify themselves as farmers, agriculture, essentially directed at local consumption, accounts for only about 37% of average household income, which is supplemented by fishing, forestry (especially in drought years) and a host of other activities.[13].Rufiji floodplain farming is therefore a traditional, labour intensive but highly efficient and low capital input system, which provided adequate livelihoods to a large resident population and, in favourable years, exported substantial surplus.

A Historical Overview of the Assessment of the Irrigation Potential of the Lower Rufiji

The approach to the development of irrigated agriculture in the Rufiji can be divided into three periods: pre-World War II (1895-1945), post-World War II (1945-1985) and post-structural adjustment (1985–present). This brief overview is largely based on the detailed historical analysis by Hoag.[14]

Pre-World War II

The upbeat 19th century reports on the productivity of the Rufiji floodplain incited the German colonial administration (1895-1917) to develop a magnificent vision of the agricultural potential of the valley. As early as 1904 the German colonial administration organised a number of expeditions inland to assess the potential of the Rufiji River for navigation, irrigation and hydropower. It was on the 1907 mission that one of the participants was killed by an elephant in the proximity of the gorge that was to be named after him: Stiegler's Gorge.[15]

The German colonial administration (and its British and post-independence Tanzanian successors) thought the valley was under-utilised and that it could be a prime production area for cash crops and especially cotton. However, the bottleneck for cotton production is the availability of labour at certain times of the year and the conflict between its crop cycle and that of traditional food crops.[16] Cotton prices need to be sufficiently high to compensate for reduced food production and high food prices.

Schematically, cotton needs to be planted when the 'long rains' rice is being harvested and needs to be harvested when the 'short rains' maize needs to be planted (Table 3). This incompatibility between 'food' crops and 'cash' crops was at the origin of the first large-scale agricultural revolt in colonial history, the Maji Maji rebellion of 1905, which started after a drought year. Three elders of the Wamatumbi symbolically pulled up cotton plants, signalling that the obligation to cultivate the cash crop was interfering with their livelihoods. The rebellion quickly spread and was brutally crushed by a war of attrition which substantially reduced the population of South-eastern Tanzania, possibly by some 30%.[17] The abandoned land to the south and east of the floodplain was colonised by vegetation and Tsetse flies. The impact of the emptying of the land is still visible today as it made possible the creation of one of the largest protected areas on the planet, the 48,000 km² Selous Game Reserve.

During the British Protectorate (1918-61) several schemes were thought up to make the Valley more productive. Telford,[18] rich with experience from the Sudan irrigation schemes, travelled widely through the Rufiji, mostly on foot.[19] He expresses some frustration with the ease with which the Rufiji people could feed themselves and really had no need for 'cash' crops. He notes that, in general, the farmers tend a miniscule plot of 0.1 ha of cotton just as a means to pay for their hut tax. He also states that, in general, there is enough rain for two crops a year and that therefore irrigation is not a necessity and can only be successful (as a supplement in bad years) if the water can be provided at a relatively small cost.

The flood hazard is, according to him, the main risk for any large - scale irrigation scheme (and also to a large - scale pumping scheme). He notes that, because of the meandering and shifting nature of the river, the distribution of the suitable soils is not continuous which would mean investing into large aqueducts, siphons and numerous cross-drainage works. He therefore concludes that 'a large scheme for irrigating the main Rufiji plains by gravity flow from a dam is unsuitable and inadvisable'. According to Telford's detailed studies, only about 40,000 to 45,000 ha could be economically developed. He thought a mechanised ploughing scheme would be most beneficial, though he warned against the maintenance costs and also put emphasis on the patchwork nature of the soils and on the numerous secondary waterways which would pose serious problems.[20] He was proved right in 1956 when the Rufiji Mechanised Cultivation Scheme collapsed after a protracted flood[21]

The views of Telford were strengthened by the analysis of Gillman:[22] 'One is therefore, forced to conclude that large-scale irrigation schemes should be left severely alone and that in the light of a recent fuller understanding of the complications of climate, soils, hydrography and markets the early optimism of the Germans regarding the possibilities of such schemes can no longer be upheld'.

Post-World War II

With the economic and political supremacy of the United States established through the proven effectiveness of its scientific and manufacturing capacity during the war and with the technocratic enthusiasm which followed the establishment of the United Nations and its specialised agencies, notably the Food and Agriculture Organisation (FAO), the approach to the Rufiji river basin underwent a fundamental change.

The pessimistic reports of the 'feet in the mud' colonial administrators and technicians were ignored as a new breed of engineers and scientists swept down on the valley, or rather started studying it from their bases in Rome or Washington. The multi-volume FAO report (1961) found 115,000 ha of irrigable land (essentially on the basis of soil type and including part of the delta on the assumption that the salt could effectively be removed). The report also stated that the Lower Rufiji could only be developed if and when a high dam would be constructed at Stiegler's Gorge. The priority areas for the development of irrigated agriculture were therefore shifted upstream to the Kilombero and Usangu floodplains.[23]

Still, decision makers have focussed ever since on the figures of the irrigable areas and have ignored the caveats that were subtly woven into the report e.g., in the conclusions of the Lower Rufiji chapter[24]: 'In wet years practically the whole of the irrigable land of the Lower Rufiji is flooded' which means irrigation infrastructure will not add any new land currently not in use.'The present population depends on the natural flooding for their rice and cotton crops and prevention of flooding would destroy their agricultural system. Irrigation in the Lower Rufiji is therefore all or nothing' which means there might be some social dimensions to deal with and there is no gradual way of accommodating this. What the people would need to do while the infrastructure was being constructed is not specified. 'Irrigation would be most appreciated in the low rainfall zone in the upper part of the valley but here the population is low. On the delta there is a high population but, with over 40 inches of rain, little demand for irrigation', which echoes Telford's[25] conclusions.

Also, the conclusion that there could be no development without a major dam stopped any further investment into the agricultural development of the Lower Rufiji. Moreover, the large floods of 1968 and 74 were used as a pretext to remove the population from the floodplain and resettle them in pilot villages of the new *Ujamaa*, the effort to restore the communal African values that, according to Nyerere, had been replaced by colonially induced individualism.[26] The dispersed habitation in the floodplain was prohibited and communal dryland farming of food crops (sorgho, cassava) and cash crops (cashew, coconut, etc.) was promoted. For the central and eastern floodplain the Ujamaa villages were in general comparatively close to the shambas so people could still continue to practice some floodplain farming after 1973 but with constraints due to time wasted in travel, etc. In the western floodplain people now had to cross the river for access to their fields, which reduced their options for combining dryland and floodplain farming. Most commonly people split their households in two during the agricultural season with women and small children predominantly attending the floodplain farm and men going back and forth. This is one of the main reasons why primary school enrolment, attendance and completion are so low in Rufiji District (less than 2% enter secondary school).

After independence a second wave of technocrats[27] started pushing the Tanzanian government for implementation of the FAO report, basically by transplanting the Tennessee Valley Authority experience.[28] In the 1970s and early 1980s a series of studies were conducted by its pendant, the Rufiji Basin Development Authority (RuBaDA), mostly funded by Norwegian aid. Their technocratic vision and lack of attention to the environmental and socioeconomic impacts of the Stiegler's Gorge dam was soon being questioned by a group of young researchers at the University of Dar es Salaam Bureau of Resource Assessment and Land Use Planning (BRALUP), as summarised in Havnevik.[29]

It is symbolic that the report by the Bureau of Reclamation for USAID[30] chose to ignore the hydrometeorological data collected by the colonial administration since the 1920s and states that the study (of the hydrology) of the Rufiji Basin starts with the FAO surveys of 1954. Sadly, the stageboards, put in place by the colonial administration and carefully monitored by local recorders several times a day, were never linked topographically to the new network of equipment established by the FAO team. The old data therefore were not used in the statistical analysis of flood height and frequency. As a result, all the engineering studies (bridge, dam) have been based on the relatively short time series of 1957 to 1984. Moreover the data set was fraught with errors.[31]

Post-Structural Adjustment (1985–present)

With structural adjustment (1985), the hydrometeorological network fell into disrepair and there were neither funds nor staff to follow up. The 1997-1998 El Niño floods, probably the largest in at least 50 years,[32] was therefore a very major surprise to the builders of the first bridge over the Rufiji (completed in 2003). As the bridge is supposed to resist the 50 and 100 year-floods with moderate to major but reparable damage, some 29 M $US had to be added to the original price to accommodate for the revised peak flow estimates. Should the Stiegler's Gorge dam have been built according to the design of the 1970s and 1980s, it would very likely have been severely damaged or even destroyed during the 1997-

1998 floods, which would have caused enormous damage and probably major loss of life downstream. Indeed, the report by Agrar- und Hydrotechnik[33] for the Rufiji Basin Development Authority points out that the blueprints for the dam developed by Hafslund of Norway did not attempt 'to route the 100 year flood through the reservoir', which they considered a design flaw. The 1997-1998 flood is estimated to have a 1 in 500 year return time,[34] therefore not only would the 3.4 billion $US dam have suffered major damage but any flood protection and/or irrigation infrastructure would have been swept away, destroying the 2.5 billion $US capital outlay.

The economics of the irrigation scheme are staggering: Assuming that the 3.4 billion $US hydropower dam would pay for itself through the sale of electricity, the investment needed for irrigated agriculture, estimated in the 1980s at 1 billion $US or currently 2.5 billion $US, would still imply 29,000 $US per ha under the, perhaps optimistic, FAO estimate of 85,000 ha or 55,000 $US per ha should the more conservative estimate of the colonial administrators and the local farmers be proved correct. The high cost is mainly explained by the need for high and solid embankments to prevent flooding of the irrigation scheme.[35]

Large-scale irrigated agriculture of the floodplain, using water from a reservoir at Stiegler's Gorge was therefore not an economically viable option. Indeed, the report states explicitly that: 'there can be no (large scale) irrigation upstream of Kipo', because of the patchy nature of the soils and this is precisely the part where rainfall tends to be deficient; even assuming double cropping, which is rarely achieved in African large - scale irrigation systems, the 'potential agricultural benefits will not be of any significant influence on the economic viability of the Stiegler's Gorge dam', even under the most optimistic scenario of irrigated surface areas, including the desalinisation of the delta; when comparing the dam with 'no dam', the effect of the dam on agricultural production in the project area 'could even be defined as insignificant'.

With such negative outcomes it is amazing that the high irrigation potential of the Lower Rufiji floodplain is still being upheld by FAO,[36] RuBaDa, the official Rufiji District development plans and in the National Irrigation Master Plan,[37] which again states that 85,000 ha of the Lower Floodplain of the Rufiji is highly suitable for irrigation.

The Linkage between High Estimates of Irrigation Potential and the Stiegler's Gorge Dam

It seems likely that the irrigation potential myth has been created and is possibly being sustained as a lobbying tool to facilitate national and donor community support for a hydropower dam by presenting it as a multipurpose dam. By raising expectations, high estimates can also be a political tool to mobilise support for high-cost development schemes. By adhering to high estimates, the concerned technical departments can attempt to increase their political clout by influencing development priorities and by raising expectations of inflow of donor funds or private investment.

Still, the attempts of the Norwegian hydropower lobby to sell the Stiegler's Gorge dam as a multipurpose dam, to accommodate for the World Bank's criticism of the project, had failed and the project was shelved. An estimated US $ 59M (US $ 24M, 1980) was spent by the Norwegian Development Aid Agency alone on the Stiegler's Gorge dam engineering studies[38] and to that would have to be added the large sums of money spent on the successive technical appraisals by other donors (FAO, USAID) and on EIAs.[39]

A Different, More Modest Vision

Alternatively, small scale irrigation, by pumping from existing secondary waterways, could possibly be developed on an estimated 4,000 ha of the Lower Floodplain. This would be especially useful in the drier western floodplain where rainfall is often inadequate. The installations would need to be mobile so they can be moved out of the floodplain prior to major floods. A thorough analysis of the social and economic aspects would be required.

The expenditure that has been committed to date on the technical studies for the large dam could easily have covered the recent building of the Rufiji bridge, the testing the viability of small-scale irrigation, the financing of a village - based land - use planning exercise for the whole of Rufiji district and even the initial implementation of those plans with the provision of microcredit.

The FAO option may be technically feasible under a very expensive scenario, but the question is whether developing such an irrigation scheme is appropriate when the current farming methods proved to be economically efficient and well -adapted to the hydrological regime of Rufiji.

Besides hydropower and irrigation, the third function of the dam would be flood control. However, floods are essential for the sustenance of the farming system, for the productivity of the natural resources on which some 150,000 people depend and also for the maintaining the biodiversity of a variety of terrestrial and aquatic ecosystems. For the local communities (the vast majority of) the floods are perceived as a blessing. In contrast, droughts are perceived as the main threat to their livelihoods. The flood risk is especially emphasised by administrators who have their origins outside of Rufiji District and was used in the 1960s and 1970s to pilot the *Ujamaa* villagisation policy. Arguably, hazardous floods could be more effectively mitigated by flood control measures in the Luwego catchment, which could possibly also benefit the wildlife that, through trophy hunting, is at present the most economical use of the southern part of the Selous Game Reserve. Some soft engineering solutions could create a series of infrastructures to slow down flow and create improved habitat for wildlife.

Conclusion

It is amazing that the myth of a vast irrigation potential in the Lower Rufiji has survived for so many decades after the thorough and pessimistic assessments by

Telford[40] and Gillman.[41] Hoag[42] has convincingly argued that, in contrast to the colonial scientists who were in close contact with the valley and its inhabitants, the post-WW II development science approach turned the valley into an abstract entity which could easily be dominated. A bottom-up development approach to small-scale irrigation supported by a microcredit scheme and targeting the driest western floodplain where food insecurity is highest is likely to be a more realistic option. It goes without saying that the recently promoted large-scale, agro-fuel development projects, in particular of sugar cane, in the same floodplain are bound to be just as hypothetical as their rice and cotton-based forebears.

Notes

1. Sunseri, T. 2003. Reinterpreting a Colonial Rebellion: Forestry and Social Control in German East Africa, 1874-1915. Environmental History. Available at http://historycooperative.press.uiuc.edu/journals/eh/8.3/sunseri.html. Duvail, S. & Hamerlynck, O. 2007. The Rufiji River Flood: Plague or Blessing? *International Journal of Biometeorology 52*. pp. 33-42.

2. Havnevik, K.J. 1993. *Tanzania: The Limits to Development from Above.* Nordiska Afrika Institutet, Sweden in cooperation with Nyota Publishers. Duvail, S. & Hamerlynck, O. 2007. The Rufiji River flood: Plague or Blessing? *International Journal of Biometeorology 52*: pp. 33-42.

3. Sandberg, A. 2004. *Institutional Challenges to the Robustness of Floodplain Agricultural Systems.* Paper presented at the Third Penannual Workshop on the Workshop Conference, Indiana University, 2-6 June 2004, available at http://www.indiana.edu/~wow3/papers/wow3_sandberg.pdf.

4. Cook, A. 1974. *A Photo-Interpretation Study of the Soils and Land Use Potential of the Lower Rufiji Basin.* BRALUP Research Paper No.34.

5. Duvail, S., Hamerlynck, O., Nandi, R.X.L., Mwambeso, P. & Elibariki, R. 2006. Participatory Mapping for Local Management of Natural Resources in Villages of the Rufiji District (Tanzania). *The Electronic Journal on Information Systems in Developing Countries 25*, 6, pp. 1-6 and 4 Figs., available at www.ejisdc.org

6. Baldus, R.D. 2004. Lion Conservation in Tanzania leads to serious human-lion conflicts, with a case study of a man-eating lion killing 35 people. Tanzania Wildlife Discussion Paper No. 41. GTZ Wildlife Programme in Tanzania, Wildlife Division, Dar es Salaam, Tanzania, available at wildlife-programme.gtz.de/wildlife/publications.html.

7. Beardall, W. 1981. Exploration of the Rufiji River under the Orders of the Sultan of Zanzibar. Proceedings of the Royal Geographical Society and Monthly Record of Geography. London. XI. pp. 641- 656. Elton, J.F. & Cotterill, H.B. 1968. 2nd Edition of Elton, J.F. 1889. Travels and Researches among Lakes and Mountains of Eastern and Central Africa. Frank Cass Publishers. *Ports au sud et au nord de Zanguebar. Centre des Archives d'Outre-Mer, Aix en Provence, France.*

8. Marsland, H. 1938. *Mlau* Cultivation in the Rufiji Valley. Tanganyika Notes and Records 5: 55-59.; Bantje, H. 1980. *Floods and Famines: A study of Food Shortages in Rufiji District.* Bureau of Resource Assessment and Land Use Planning, Research Paper N° 63. University of Dar es Salaam, Tanzania. Sandberg, A. 2004. Institutional Challenges to the Robustness of Flood Plain Agricultural Systems. Paper presented at the Third Penannual Workshop on the Workshop Conference, Indiana University, 2-6 June 2004, available at http://www.indiana.edu/~wow3/papers/wow3_sandberg.pdf

9. Duvail, S. & Hamerlynck, O. 2007. The Rufiji River Flood: Plague or Blessing? *International Journal of Biometeorology 52*, pp. 33-42.

10. Doody, K.Z., John, P., Mhina, F. & Hamerlynck, O. 2003. Merging Traditional and Scientific Knowledge for Environmental Awareness. The World Wetlands Day celebrations held in Utete,

Rufiji on February 2, 2003. IUCN REMP Technical Reports N° 33, available at www.iucn.org/themes/wetlands/33wwd03.pdf

11. Sandberg, A. 2004. *Institutional Challenges to the Robustness of Floodplain Agricultural Systems.* Paper presented at the Third Penannual Workshop on the Workshop Conference, Indiana University, 2-6 June 2004, available at http://www.indiana.edu/~wow3/papers/wow3_sandberg.pdf

12. Bantje, H. 1980. *Floods and Famines: A study of Food Shortages in Rufiji District. Bureau of Resource Assessment and Land Use Planning,* Research Paper N° 63. University of Dar es Salaam, Tanzania. Havnevik, K.J. 1993. *Tanzania: The Limits to Development from Above.* Nordiska Afrikainstitutet Sweden in cooperation with Nyota Publishers, Tanzania.

13. Turpie, J. 2000. *The Use and Value of Natural Resources of the Rufiji Floodplain and Delta, Tanzania.* IUCN REMP Technical Reports N° 17. Sandberg, A. 2004. *Institutional Challenges to the Robustness of Floodplain Agricultural Systems.* Paper presented at the Third Penannual Workshop on the Workshop Conference, Indiana University, 2-6 June 2004, available at http://www.indiana.edu/~wow3/papers/wow3_sandberg.pdf

14. Hoag, H.J. (2003, PhD thesis). *Designing the Delta: A History of Water and Development in the Lower Rufiji River Basin, Tanzania, 1945-1985.* Boston University, USA.

15. Hoag, H.J. (2003, PhD thesis). *Designing the Delta: A History of Water and Development in the Lower Rufiji River Basin, Tanzania, 1945-1985.*

16. Yoshida, 1977. Agricultural Survey of the Lower Rufiji Plain. Japan Association for African Studies, Kyoto, Japan. *Journal of African Studies 16,* pp. 36-58.

17. Pakenham, T. 1991. *The Scramble for Africa : White Man's Conquest of the Dark Continent from 1876 - 1912.* Random House.

18. Telford, A.M. 1929. *Report on the Development of the Rufiji and Kilombero Valley.* Crown Agents for the Colonies, London, U.K.

19. Hoag, H.J. (2003, PhD thesis). *Designing the Delta: A History of Water and Development in the Lower Rufiji River Basin, Tanzania, 1945-1985.*

20. Telford, A.M. 1929. *Report on the Development of the Rufiji and Kilombero Valley.* Crown Agents for the Colonies, London, U.K.

21. Hoag, H.J. (2003, PhD thesis). *Designing the Delta: A History of Water and Development in the Lower Rufiji River Basin, Tanzania, 1945-1985.*

22. Gillman, C. 1945. *A Reconnaissance of the Hydrology of Tanganyika Territory in its Geographical Settings.* Water Consultant's Report N° 6. [Government Printer] Dar es Salaam, Tanzania.

23. Franks, T., Lankford, B. & Mdemu, M. 2004. Managing Water amongst Competing Uses: The Usangu Wetland in Tanzania. *Irrigation and Drainage 53,* 3: pp. 277-286.

24. FAO. 1997. Irrigation Potential in Africa: A Basin Approach. *FAO Land and Water Bulletin 4.* Rome, Italy.

25. Telford, A.M. 1929. *Report on the Development of the Rufiji and Kilombero Valley.* Crown Agents for the Colonies, London, U.K.

26. Hyden, G. 1980. *Beyond Ujamaa in Tanzania: Underdevelopment and an Uncaptured Peasantry.* Berkeley: University of California Press.

27. USAID. 1967. *Rufiji Basin: Land and Water Resource Development Potential.* Bureau of Reclamation, United States Agency for International Development; Department of State, Idaho.

28. Hoag, H.J. (2003, PhD thesis). *Designing the Delta: A History of Water and Development in the Lower Rufiji River Basin, Tanzania, 1945-1985.*

29. Havnevik, K.J. 1993. *Tanzania: The Limits to Development from Above.* Nordiska Afrika Institutet, Sweden in cooperation with Nyota Publishers.

30. USAID. 1967. *Rufiji Basin: Land and Water Resource Development Potential.* Bureau of Reclamation, United States Agency for International Development; Department of State, Idaho.

31. JBG Gauff Ingenieure. 2000. *Rufiji Bridge and its Floodplain Crossing: Supplementary Hydrological Study and Evaluation of Flood Records following the 1997–1998 El Nino Floods.* Final report for the Ministry of Works, United Republic of Tanzania. JBG Gauff Ingenieure, Frankfurt.

32. Erftemeijer, P.L.A. & Hamerlynck, O. 2005. Die-back of the Mangrove *Heritiera littoralis* in the Rufiji Delta (Tanzania) following El Niño floods. *Journal of Coastal Research 42.* pp. 228-235.

33. Agrar- und Hydrotechnik. 1982. Irrigated Agriculture in the Lower Rufiji Valley. Pre-feasibility Study. Essen, Germany.

34. JBG Gauff Ingenieure. 2000. *Rufiji Bridge and its Floodplain Crossing: Supplementary Hydrological Study and Evaluation of Flood Records following the 1997–1998 El Nino Floods.* Final report for the Ministry of Works, United Republic of Tanzania. JBG Gauff Ingenieure, Frankfurt.

35. Agrar- und Hydrotechnik. 1982. Irrigated Agriculture in the Lower Rufiji Valley. Pre-feasibility Study. Essen, Germany.

36. FAO. 1997. Irrigation Potential in Africa: A Basin Approach. *FAO Land and Water Bulletin 4.* Rome, Italy.

37. Japan International Cooperation Agency (JICA). 2001. *The Study on the National Irrigation Master Plan in the United Republic of Tanzania.* Ministry of Agriculture and Food Security.

38. Bryceson, I. 1988. *Anticipated impacts of a Norwegian-proposed Dam at Stiegler's Gorge, Tanzania.* Paper presented at 'Hearing on Third World Environmental Perspectives' on the Brundtland Commission Report, Oslo, November 1988, available at www.fivas.org/pub/power_c/k14.htm

39. Euroconsult & Delft Hydraulics Laboratory. 1980. *Identification Study on the Ecological Impacts of the Stiegler's Gorge Power and Flood Control Project.* Dar es Salaam, Tanzania. FAO. 1961. *The Rufiji Basin, Tanganyika.* Report to the Government of Tanganyika on the Preliminary Reconnaissance Survey of the Rufiji Basin. Rome, Italy.

40. Telford, A.M. 1929. *Report on the Development of the Rufiji and Kilombero Valley.* Crown Agents for the Colonies, London, U.K.

41. Gillman, C. 1945. *A Reconnaissance of the Hydrology of Tanganyika Territory in its Geographical Settings.* Water Consultant's Report N° 6. [Government Printer] Dar es Salaam, Tanzania.

42. Hoag, H.J. (2003, PhD thesis). *Designing the Delta: A History of Water and Development in the Lower Rufiji River Basin, Tanzania, 1945-1985.*

Irrigation among the Chagga in Kilimanjaro, Tanzania: The Organisation of Mfongo Irrigation

Mattias Tagseth

Introduction

The southern slopes of Mt. Kilimanjaro, the world's largest free-standing mountain and rising to 5985 m in north-east Tanzania, constitute a green and moist oasis surrounded by a semi-arid low plain. From the early Iron Age[1] onwards, the mountain provided refuge from drought and vectors of disease, opportunities of fertile soils and water as well as new challenges such as cold and cloud cover to groups of people that came to settle and cultivate there. Over the last several centuries the groups that are now known collectively as the Chagga were formed. Most of these settlers appear to have been Bantu-speakers from the surrounding mountains and plains, but there have been influxes from other groups. The management practices developed or adopted here to suit the conditions on the mountain slope are now a part of the cultural and technological heritage of the Mt. Kilimanjaro region. While experiencing challenges of modernisation and increasing pressure on resources, smallholders continue to rely on the *mfongo,*[2] which is a gravity-fed canal system, and on the *kihamba,* a local form of agroforestry in combination with other pursuits.

Gravity-based canal irrigation is a longstanding tradition on the southern slopes of Mt. Kilimanjaro, and an extensive network of irrigation canals remains in use. This region saw rapid social and economic change during the 20th century, with associated changes in water use and social organisation. Technical and organisational development in indigenous irrigation has followed, but *mfongo* irrigation has escaped previous attempts at incorporation in the national schemes for water management, and instead irrigation extension has been focused on a small number of lowland schemes and the Lower Moshi Irrigation Scheme. An extensive irrigation infrastructure on the southern slopes of Mt. Kilimanjaro is documented by contemporary 19th century reports. Today, this area probably has the largest cluster of 'hill furrows' in eastern Africa. Farmer managed hill furrow systems with some similarity in organisation and technology are found for instance along the Rift valley[3] in Tanzania and Kenya, in nearby mountain areas as well as some locations in the plains surrounding Mt. Kilimanjaro.[4] These were grouped by Adams[5] into the Kerio, the Sonjo and the Kilimanjaro clusters.

Several aspects of 'indigenous irrigation'[6] are of interest for current research. The expansion of 'hill furrow' irrigation in Kilimanjaro has been addressed elsewhere.[7] The meeting of modern river basin planning with traditional models of water management from the 1990s led to conflict and a breakdown in communication, which is still a central challenge in the management of the Pangani River Basin. In this situation there is a need for increased knowledge of practices of water use and water management in order to guide sensible intervention in local water management.[8] The farmer-managed irrigation sector is much larger in size than the engineered sector of irrigation and the commercial estate sector combined. Tanzania is a country troubled by poverty, inadequate food security and a poor level of human development[9]. Irrigation development has been devised as one remedy for these problems, and is on the political agenda.[10] The development of new and capital-intensive irrigation schemes on a larger scale is seen as problematic, due to high costs in relation to the benefits and competition over water for other economic uses, among other factors. The redevelopment of the whole 'traditional irrigation' sector, viewed by actors in irrigation extension as deficient both in terms of technology and organisation, has thus been devised as central to the strategy to improve food security by the Ministry of Agriculture.[11]

The strategy of traditional irrigation rehabilitation is developed out of criteria of 'water use efficiency', and the perception that these are imperfect irrigation schemes which can be upgraded to a higher 'irrigation development level' through technology transfer at a comparatively lower cost.[12] With 1189 schemes and more than 500,000 ha, representing 77% of the irrigated acreage according to the inventory of the Ministry of Agriculture and Food Security,[13] traditional irrigation represents both a substantial production and a potential for improvement. Some caution is warranted, however, since the assessment is not based on an evaluation of the actual experience of traditional irrigation rehabilitation which has taken place from the early 1980s onwards. This strategy of rapid redevelopment aims to replace both the technical infrastructure and the established organisation of the schemes, and renders improved knowledge of the organisation, operation and sustainability of farmer - managed irrigation in Tanzania an urgent matter.

Theory of irrigation systems management has been dominated by organic metaphors, reflecting a functionalist influence, as exemplified by the influential models where the sustainability of an irrigation scheme is assessed through the analysis of how its 'functional needs' are served.[14] Normative aspects are also emphasised in the theme of governance from Common Property Resources theory, which seeks to analyse and evaluate the commons through the properties of their institutions and especially in terms of their formal and operational rules.[15] Diachronic studies of environmental change,[16] but also a critique of applications of equilibrium theory have been important in challenging hegemonic misconceptions of the African environment and in resource management.[17] While useful in addressing questions of social control, theories commonly used in the analysis of irrigation organisation can also constrain analysis due to the inherent tendency to describe organisations as static and apolitical objects.[18]

There is thus a rationale to look for supplementing modes of analysis, for instance in order to understand change or conflict in irrigation systems and water management. According to Vincent[19] the third main approach in the literature on the management of canal systems focuses on hydraulic tenure. Some alternatives can be found in studies focusing on the social organisation of water use drawing on processual anthropology.[20] Furthermore, studies of negotiation (conflict, conflict resolution) over water[21] and other resources[22] which draw on concepts of legal pluralism are likely to give results differing from those starting with the problems of control. The debate on land tenure in African studies is an important and much more developed parallel to that of water tenure. Land tenure institutions involve complex bundles of rights, subject to change and the arenas of land allocation are only partially isolated from other arenas.

The objective of this article is to describe and analyse the social organisation of *mfongo* irrigation on the slopes of Mt. Kilimanjaro on the basis of a case study. The focus is some aspects of the social organisation of irrigation: firstly, on scheme tenure, water tenure and access to water, and secondly, on the canal organisation and its relationship to local organisation. This article relies on the principles of methodological triangulation[23]. The analysis of irrigation organisation will be done through the comparison of a normative (institutional, symbolic) level with reports on the social practice of water management at scheme level, and with the result of this process in terms of water access for different groups, while drawing on quantitative and qualitative data.

Mt. Kilimanjaro, Tanzania

Photo: Google images

Water and the Chagga of Mt. Kilimanjaro

Mt. Kilimanjaro is a diverse region, and a research strategy focusing on the central District of Moshi Rural on the southern slopes was chosen. In this article I draw on the results of fieldwork conducted in the period 1995–1996, supplemented during visits in 2000–2002, in a region within this district, and also on data from one mountain village selected for a more intensive study. The catchment of Himo River was selected as a regional study area. Centrally located in this hydrological region is the district of Marangu, which literally translates as '*a lot of water*'. Water use by the Chagga of Kilimanjaro and some dimensions of variation within that system will be introduced, before proceeding with a description of upland canal structures, based on a case study of irrigation in an upland village. The case study draws on a series of interviews in all the 65 households which have physical access to four irrigation schemes, field conversations, depth interviews and observation. The establishment and use of the system is discussed before moving on to the institutions, norms and symbolism surrounding the system, some data on differences in access to water between social groups before presenting reports on negotiations and, finally, the concluding remarks.

The extensive network of small-scale irrigation canals on the southern slopes of Mt. Kilimanjaro was marvelled upon by the early European visitors who first reported on them in the 19th century.[24] Oral history and written accounts from the study area record that hydraulic control was well established prior to contacts with Europe. It is clear that the area has a long history of agriculture. While continuity of settlement is less certain, the theory forwarded on the basis of archaeological survey is that sites were first settled by farming and iron-using communities in the first millennium.[25] The irrigation technology in Marangu can be traced back to the first half of the 19th century for specific schemes while about three and a half centuries are indicated in more general oral traditions.[26] This is as far as oral history can be expected to take us, while systematic historical landscape studies and archaeology might yield more knowledge. Moshi and the neighbouring Hai Districts have *c.* 1000 irrigation schemes of varying size, with an average throughput of 18.7 litres per second.[27] Some of these systems have main canals serving several hundred farms 6–10 km from the source, while most are less than 2 km in length and serve from 10 to 30 farms. In addition to the local traditions of water use by *mfongo*, the combination of the banana-based agroforestry grove (*kihamba*) and the arable (*shamba, pata*) cultivation remains an important feature of the farming system.

In the 19th century, the mountain was divided between chiefdoms or counties that had changing alliances[28], and were competing for control over the Swahili trade.[29] Each more-or-less sovereign chiefdom (*nchi*) was organised into several districts (*mitaa*) under a headman (*mchili*). Patrilineal kinship and a preference for virilocal settlement led to the development of the localised patrilineage as an important arena and group with a territorial interest.[30] There are no compact village settlements, but a densely settled landscape where each small homestead farm borders on the next, and where there is a high likelihood that your neighbour is also your agnate. The *mangi* (Prince) could be involved in the construction of a scheme[31] and water management[32], but most furrows would have been started by prominent members of smaller local communities. In the early 20th century,

Gutmann[33] distinguished between several types of canal tenure. Some schemes were started by specialists,[34] and some schemes were under the management of the *mangi,* as a part of the family estate or for other reasons, but the majority were under the control of lineage elders.

A preoccupation with water and its integration into Chagga culture is reflected in beliefs, proverbs, riddles and sayings, such as the blessing *'may you sleep below a furrow'*, which is still heard. The value of irrigation in food security is communicated in riddles, as shown in an example provided by Mosha,[35] structured with a question, *'My father has left me a bowl from which I have been eating ever since'*, and an appropriate answer, *'The irrigation canal'*. Aspects of a gendered symbolism of irrigation[36] is seen in the current prohibitions against women going to the intake of the furrow or working on the canal, and in reports of how in the past pregnant women could not cross the furrows.[37] Scheme 'ownership' often lies with the descendants of the initiator of a scheme, and it carries the right to manage it. In some areas, people say that a man who starts a new furrow will benefit, but that he will not reach a high age. Some furrow schemes are regarded as enchanted, which serves to strengthen norms of behaviour, tenure and rights of use in a water commons. Arguments that a goat should be paid for and slaughtered in order to restore the flow of water and by way of compensation have appeared both in conflicts over tenure related to rehabilitation and in conflicts with the government officers over access to the intake, especially during a campaign to strengthen the statutory water management system from the early 1990s.

Dimensions of Variation

The topography of the area is dominated by the slope of the huge volcanic cone, which is intersected by alternating ridges and river valleys. Important differences in the resource base are determined by the variations in climate, from the cool sub-humid highlands to the hot semi-arid foot-lands. The vertical linkages and the river valleys as barriers to horizontal communication are important in the spatial structure. Data from the catchment of the Himo River[38] suggest that the *mfongo* canals have been established in different periods in different areas. The mid-slope core area has the longest history of settlement, and the some of the older schemes are to be found here. Further up, the establishment of some schemes can be dated through genealogies to the first half of the 19th century. Some schemes in the lowlands are also pre-colonial, but furrow development continued here in the German colonial and early British periods. Scheme establishment continued with migration throughout the 20th century, and settler schemes have been revitalised with migration to lowland areas since the period following villagisation and nationalisation of estate land in the 1970s. Thus, there has been an apparent association between the establishment of *Mfongo* schemes and settlement history over a long time period.

The methods of irrigation are predominantly extensive. Supplementary and protective irrigation makes food production less vulnerable to drought. Permanent crops are irrigated in locations that are marginal for rain-fed banana and coffee cultivation. The irrigation of arable crops remains important. In the higher areas, maize is cultivated in the relatively dry period following the main rains, whereas further down slope, irrigation enables a second cultivation period during the period of short rains. In the foot-lands, arable cultivation is somewhat marginal even during the main rains, and is safeguarded by supplementary irrigation. The need to irrigate finger millet (*mbege*) was important in the establishment of many canals. This motivation can be attributed to a past phase of the expansion of irrigation, predating the expansion of coffee cultivation and an associated transformation of *kihamba* agroforestry practices, which took place in the first half of the 20th century. In the moist higher areas, millet cultivation took place in the period following the main rains (*masika*).

This crop has been uncommon for more than 60 years, as arable fields were converted into *kihamba* groves, while the remaining arable was turned into maize fields. Millet could provide high yields,[39] but required more labour to produce than maize. The exploitation of the pre-existing water supply for new purposes slowed down the rate of decline in the technology. Horticulture on a commercial scale has been increasing in recent decades. Current water use differs according to purpose and season along the slope. In good years, the demand for irrigation is limited during the rainy seasons in the uplands, while these periods are important for arable cultivation safeguarded by supplementary irrigation in the plains and the foothills. The different timing of the peak water demand between three zones along the slope reduces the regional competition for water somewhat. Farmers who cultivate below lowland furrows are more likely to report irrigation than those in the upland area. The main purposes of irrigation also differ, and water use is more intensive in the communities on the lower slope.

There is a significant awareness of the varying natural conditions along the slope, the possibilities and restrictions that they pose. There is also variation in the history and institutional arrangements between different areas, and from one commons to the next. Correspondingly, people do not expect the same norms and regulations of irrigation to apply everywhere. When irrigators in a lowland case study area learned that I was researching simultaneously in an upland area, they were curious to know whether goats were still sacrificed there. In the upland area, irrigators contrasted their irrigation practices with those '*in the porini*' (lowland bush), where they have to bribe water distributors or bring their bush-knives in order to ensure water allocations. Irrigation has been practised for generations in the lowland areas too, but these areas are marginal in the pre-colonial Chagga sphere of influence. In the uplands these areas were spoken of as having no tradition of irrigation. I have previously[40] offered a categorisation of farmer-managed irrigation organisation into two main modes:

• Furrow organisations of the first type are centred on unnamed and informal groups of users. The authority over the scheme rests with a furrow chairman,[41] and depends on his relationship to the canal founder. The canal rites and the mystical ideas surrounding the schemes have been weakened as a consequence of school attendance and conversion to Christianity among other things, but they remain an influence on irrigation management. The furrow chairman is appointed by his patrilineage or 'clan', and usually holds his office for life. The furrow chairman is often said to own the canal, but the resulting clan control over the water is limited. The main responsibility of the furrow elder is to organise the maintenance of the scheme, and to call out work parties when necessary. Certain canal rights are defined by kinship, but membership in the group of furrow users is defined by territory (i.e. command area) and contribution to scheme maintenance rather than by kinship. The canal organisation is an unnamed and informal social group with members (households), who at least in local theory invest labour in the scheme and have more or less equal claims to water. Group membership is regulated by the ability to access land[42] in the command area and participation in the work party. The furrow elder controls changes in scheme layout and the composition of the work party. The water itself is understood to be *'a gift from god'*, and cannot be claimed as property by anyone. To demand payment for water is thus regarded as *'a sin'*. The access to water depends on the use of the canal, and this is governed by principles of contribution to the scheme: *'The water belongs to those who follow it to the source'* it is said. A schedule of rotation could be agreed upon at a general meeting in case of drought, and this appears to be the regular practice in some areas. Under normal conditions, however, the allocation of water in upland Marangu was a matter of direct negotiation between co-users.

• The second type of furrow organisation has a higher degree of formal regulation and the leadership is in the form of an elected or appointed board and canal leader. While the rules governing contribution to the scheme and the allocation of water are more formal, the details vary significantly from scheme to scheme. Under this mode of organisation, the allocation of water is commonly divided into fixed time units per user, whereas this is less usual under the first mode. The rotation between secondary branches or geographical areas is also common. In this second mode of irrigation organisation, membership fees, fines for the violation of the norms of irrigation and – according some users – bribes are used in the regulation of access to the scheme and its water. A connection between cash and water is less problematic here. The group of users can be formalised as part of the village organisation where the scheme does not transcend the village, or separately, for instance in the form of a co-operative. There is a stronger precedence for statutory water rights. This is, firstly, because this mode of organisation is often found on land nationalised or purchased from the former estates, and, secondly, due to some influence through government assistance to these irrigation schemes.

The first mode of organisation predominates in the upland schemes, which are older and usually smaller than the schemes further down the mountain slope. The mid-slope situation is varied, in contrast to the foot-lands where organisation according to the second mode of furrow organisation dominates. The background and scale of the schemes also vary. There are different problems of water management associated with these two modes, though it is apparent that management according to both principles can be relatively successful at scheme level. A common trait of the farmer-managed irrigation in the area is that the institutions of irrigation are well - integrated with other local institutions. Universal user rights based on access to land in a command area subject to contribution to the scheme are acknowledged. The definitions of rights thus serve to improve the sustainability of the labour-intensive schemes.

In the schemes described in the following sections, the use and management of water under the first (or lineage) mode, is described using the case of an upland village[43] in Marangu. The village stretches out from the upper fringe to the core area of the Chagga settlement, and lies within the highest ecological zone.

Upland Canal Structures

There are differences in the density of schemes and how these have survived within short distances of each other. In areas closer to the main road and the central place in Marangu there are a lot of abandoned schemes, while other areas (including the village selected as a case study area) have a high density of active irrigation canals. All the active and derelict schemes have probably never operated simultaneously. The springs, natural waterways and canals in this study area are so densely located, that it can be difficult to find anywhere where the sound of trickling of running water cannot be heard. The area has two sources of furrow water, the Ruwa stream and its tributary Kizuru stream. Both streams are spring - fed, and supply a number of furrows. When I climbed the waterways of this small sub-system, I identified 36 individually named furrows. One-third of them took their water from the Kizuru stream. Of these, some 17 furrows deliver water to different parts of the case study village.[44] The furrow network is extensive, and it has a high connectivity. There are approximately 30 households for each furrow scheme, as compared to 50 households for each public standpipe. The schemes are comparatively small, with actual discharges ranging from 8 to 42 litres per second.[45]

The stream water is checked, either by a small permeable dam or by a stonewall protruding into the streambed, in order to create some hydraulic head. This dam is constructed with boulders, stones, twigs, and mud. The intake is often constructed where the water flows more smoothly, and where the stream divides naturally or at a pond. Permeable structures and overflow structures are useful, as the stream flow is variable. Both high flow and low flow make repairs or modifications of the check structure necessary from time to time. In many cases, a furrow will divert most of the available water during the dry parts of the year. However, some water is always available for the next furrow below, since water is either lost or returned close to the abstraction point, and because stream flow is augmented by the seepage of groundwater infiltrated on higher ground. The technology, with the use of permeable dams, limits the maximum water acquisition at a single abstraction point.

A canal usually starts from a source area in a gorge or depression and also collects water on its way to the farms on the ridge. Simple spillway- or overflow structures are incorporated close to the intake in order to avoid breakdown of the canal banks at times of high flow. In the river valleys, the water is carried on a gentle gradient in a canal running along an embankment, in places several metres high, towards the higher ground on the ridge.

Regular abstraction points in the earth banks of the canal can be seen from the small packs of grass and mud used to block them. Upon reaching the main settlement on the ridge, the furrows divide. The proportional weir is used, as permanently flowing water is wanted in the secondary canals, which serve the homestead compound farms. This is different from areas with less water, where water is rotated according to a sequence agreed upon at a general meeting following the communal work session, or divided proportionally into a schedule of time-units. The distribution of water is manipulated by placing and adjusting a small pack of mud, twigs and grass on either branch. The canals divide to provide running water close to most homesteads and fields. Other secondary canals are dry. The terrain on the ridge is sloping, and it is sometimes necessary to bring the water down rapidly, which requires a controlled release of potential energy.

The simple drop structure used in order to achieve this also provides a convenient tapping place for domestic water. Stone lining is another method of energy control, which is used to prevent scouring. The horizontal footpaths tend to run alongside the canals in the river valleys and on the ridge. The co-localisation of paths and canals has the advantage of facilitating the monitoring of the canal and the flow of water. The situation of turning points on the public paths rather than in private groves further facilitates the monitoring of water flow. The control of runoff is more important than the drainage of irrigation water in the sloping and naturally well-drained terrain. Surface water is drained into the river through the irrigation canals and separate drainage canals.

The Establishment of the Upland Furrows

The oral history of the *Mfongo* schemes in the upland case study area suggests that the establishment of the schemes was associated with the immigration and subsequent expansion of the furrow-controlling lineages in the area, more than with development from subsequent intensification of agriculture. Scheme establishment in this village is estimated to have taken place in the period from between the 1830s and 1920s, when the margin towards the forest reserve was reached. Even though '*no one would remember how to survey a furrow today*' (Informant B), existing schemes in this area continue to be maintained and used. The first simple structures were made with digging sticks. They were later expanded to include more farms, and improved by subsequent generations. Oral tradition suggests that there were two centres in Marangu prior to the arrival of the ancestors of the lineage of the chiefs (Ngowi/Lyimo/Marealle). [46] Engraved stones related to the boys' initiation (*ngasi*) ritual have been reported at two locations in Marangu, [47] one of which is located near the hamlet of *Kisu* (the knife). This and a place-name which might be associated to ritual sites imply that the area is long established cultural space. Lack of indications in archaeological survey by Odner, [48] interview data and the village name of *Komalyangoe* (where the poles grow) do not suggest that this particular area was among the early agricultural settlements within the wider upland area. In this regard, the history of the schemes of one lineage can serve as an example. [49]

The furrow of Ndorano was the first of a series of small furrows built by the members of this family as they grew in numbers. The history of this family in the area started with a settler who migrated here due to conflicts over pasture between two clans, and the first furrow was initiated by his son (no. 2). It is named after a former furrow chairman. The furrow chairman explained: '*The name changes for each generation, it is the grandfather furrow*'. The furrow was constructed '*because it was too far to go to the river to fetch water for irrigation and domestic purposes*' (Informant E). In the third generation, the resident lineage continued through sons no. 3 and no. 6. The last born son (no. 6) took over his father's homestead, while his older brother (no. 3) had other plots in the area. In the fourth generation, the son (no. 7) of no. 3, built the second furrow in order to provide water for irrigation of finger millet, maize, beans and banana in an area above the original furrow, and he set up a homestead there. Water was also used for domestic purposes and for fire protection. This second furrow is called *the furrow of Saranga*, after a former chairman.

According to the current chairman (Informant D), '*There was no involvement from the mangi. The people of the area used to organise themselves to do it. All the people who were going to use the water took part.*' Also, in the fourth generation, the son (no. 10) of the last born (no. 6) constructed the third and last furrow, the smallish furrow of *Kisipio*. The scheme improved the water supply to the former tail-end users of the first furrow, as new farms were established. According to the current furrow chairman (Informant E), who is the great grandson of the furrow initiator, it was constructed for '*irrigation of finger millet. There is too little land for that now; we mainly irrigate vegetables and beans*'.

The detailed genealogy shows that the establishment of the schemes was related to the expansion of the resident lineage. Control over the furrow has been inherited by the direct descendants of the canal founders, despite the inclusion of non-kin in the group of users though land acquisition and migration. There is a correspondence between the three segments of the lineage and control over the three schemes, but there is no apparent correspondence between membership in a lineage segment and the use of a scheme under the control of that particular segment. The family history describes a process, where three canals are a physical result of the past expansion of a lineage. The addition of the second furrow (*Saranga*) increased the command area of furrows and the volume of water under the control of the family, but may also have served to provide independent access for an emerging lineage segment. The third scheme (*Kisipio*) may have eased the supply for areas which it was difficult to acquire water for due to competition with upstream users.

Springs understood to be controlled by ancestors who died or 'disappeared' there are common on Mt. Kilimanjaro and nearby areas. One interpretation of the sacred springs and rain shrines that are common in the mountain landscapes of north-east Tanzania is that they are markers of a symbolic political geography, involved in the definition of groups of people, tenure, and political regions or domains. The clan of the Prince of Marangu is associated with a central hilltop shrine, the 'ancestor hill' where the ancestral mother of the Marealle, Ngowi and Lyimo clans stayed and died following the migration from *Ukambani* to the north of Mt. Kilimanjaro.

In times of famine and drought, '*they went to the Fumwu lya Meku hill above and slaughtered a bull and a male goat. It used to start raining the same day*' (Informant M). Tradition holds that a named child and later a man were killed at a spring in the upland part of the case study area, and this is still respected as a sacred place (Informants H, I and P). It is said that at the time of scheme construction, and in later times of crisis, a goat or a bull was sacrificed there in order to make the water flow smoothly (Informants G and B). In the former case, it was made to the supreme god (Ruwa), while in the latter case it was to the ancestors. One furrow chairman reported that he offered libations of milk and *mbege* beer for the same purpose.

Libations in connection with the furrows have declined, and are done discretely. Elders complain that the discharge of the springs and the rivers has diminished, which can be related to physical processes and discourses of climate change, deforestation and land use changes which lead to less infiltration of water, but also to a local discourse of moral decay. Common to the furrow intake, the sacred spring and the hilltop shrine is that they all have to be manipulated by the 'proper' groups of people and procedure, which is important in their use as vehicles for the establishment of political control at different levels through devices of 'enchantment'. Conflicts involving references to the symbolism of an irrigation scheme or a water source can thus relate not only to the control over water as a resource or the 'mystification' of hydraulic tenure, but also to local discourses of power.[50]

Water Use

The importance of irrigation has declined with changes in the upland agricultural system, primarily the extension of the area under permanent crops: '*In the past, there were open areas, where people could plant more seasonal crops. These needed irrigation. These days, there are a lot of permanent crops covering almost all the open area. ... People are not as interested as they used to be, but we still regard irrigation as important'* (Informant K). Coffee is processed on the farms by the wet method, which requires access to water. The cherries are pulped in running water (reported by 83% n=65), before the beans are fermented in basins, and then washed and dried for sale in parchment. This method ensures the high quality of the acidic *Arabica* coffee from Kilimanjaro. The farmers struggle to keep the mole rats (*mfuko*) away from crops, and flooding their tunnels was reported by 26% of the farmers. Livestock, mainly zebu, are fed with succulent banana stems and rarely need to be given water (8%). Domestic use includes water for drinking (25%), cooking (60%) and washing (94%).

Irrigation was reported by 60% of the households with land in the command area, and can be described as supplementary for most crops. For some crops the access to water acts mainly as a safeguard against crop failure. Irrigation is used in order to extend the cultivation period into parts of the year when there is a higher risk of drought. Maize and beans are watered in September and October. Some farmers also water the coffee trees at this time or in the dry spell in January, together with other *kihamba* crops[51]. For crops such as sweet taro, sugar cane and watercress, there is a continuous need for water outside the rainy seasons. The average yields of coffee, beans and maize are higher in households where irrigation is practised (p> 0.95, independent samples T-test). The incidence of vegetable cultivation is also affected by irrigation practice (p>0.95, chi-square test).

Households with land in the command area were differentiated by key informants, in three independent sessions by wealth rank and into three broad socio-economic groups. The data sets were evaluated through rank correlation and reduced to a trichotomous variable. Variation in water use between social groups can be shown when irrigation activities are further differentiated into three broad categories: arable, agroforestry and horticulture (Table 1). The middle group has a higher frequency of reported irrigation, especially for arable crops (dry season maize). Vegetable cultivation is on a small scale, for household consumption and the local market. Horticulture and bananas are the main source of cash for women.

TABLE 1. REPORTED IRRIGATION BY WEALTH RANK

Type of irrigation	Well-off 27% (17)	Middle 48% (30)	Poor 25% (16)	All groups 100% (63)
Arable	29.4% (5)	46.7% (14)	25.0% (4)	36.5% (23)
Agroforestry	23.4% (4)	6.7% (2)	18.8% (3)	14.3% (9)
Horticulture	52.9% (9)	40.0% (12)	37.5% (6)	42.9% (27)
All purposes	58.8% (10)	63.3% (19)	56.3% (9)	60.3% (38)

Source: Author's Compilation

Local Arenas, Groups and Leadership

Processes of social control, conflict and co-operation in resource use take place in a local arena. The village organisation, the churches and the coffee co-operative can be understood as formalised and 'modern' organisations. Other groups, organisational units and sets of statuses can, for the sake of convenience, be labelled 'traditional'. The latter are important in the organisation of agriculture and irrigation.

The smaller neighbourhood convenes from time to time in order to discuss common problems. Small parleys are conducted in conflict resolution; where the complainant, the 10-cell leader (below) and an elder from the neighbourhood are present. The co-users of a furrow are also neighbours and relatives. They exchange surplus crops and favours, and borrow the use of auxiliary plots from each other. Neighbours and relatives have obligations of reciprocity. These networks are crucial in times of crisis and for resource mobilisation, and community members monitor each other closely. Interdependency also provides a means for sanctioning. Loss of standing could influence the future allocation of a vacant lineage plot, the lending of a cow, or the willingness to contribute towards the cost of a marriage or medical treatment. Violence does occasionally occur, but the ultimate sanction is to be ostracised: to be expelled from the community,[52] deprived of the means of subsistence and cut off from the web of mutual assistance.

Every Chagga belongs to one of the more than 400 major patrilineages of the mountain, each named after a founder. The precise relationship to the founder is unknown, so these *ukoo* can be defined as clans.[53] This clan is exogamous, and it can share food prohibitions but it has no joint assets and it does not act as a body. The active group is the smaller virilocal lineage segment, the *localised patrilineage*.[54] This is a group of men with a common ancestor and history in the area, together with their dependants. At times enforced by neighbours, the kin groups appear as tiers or segments, as the beer-drinking group (lineage members and neighbours), the cattle-slaughtering group (approximately 20 married men) and the goat-slaughtering group (an elder and his married sons). The localised patrilineage is a resilient organisational unit, which remains important in the management of the furrows, and matters of mutual assistance, marriage and land.

Modern institutions of governance were introduced into the area after independence, as part of the experiment with *ujamaa* or African socialism, replacing the 'native authority', of indirect rule which was disposed of as a colonial creation. The villagisation programmes of the mid-1970s involved movement of people on a grand scale and the creation of new settlement patterns in most parts of Tanzania. In the established communities on the slopes of Mt Kilimanjaro, however, old territorial units were reshuffled into new agglomerates and realities changed less.[55] A village is headed by a council, which elects a chairman and establishes a series of village committees. Some village administrations are involved in the matters of the furrows through the village committee of production. The next tier in the structure is the hamlet (*kitingoji, mtaa*), headed by a chairman. These chairmen are more frequently involved in matters of conflict resolution. In the case of *Komalyangoe Kisu*, the roles of hamlet chairman, head of a lineage and chairman for one of the furrows were combined in one person. A system of a cell leader (*balozi*) for 10 households was introduced on Mt. Kilimanjaro in the mid-1960s.[56] Interviews in four village offices yielded no information on involvement in matters of the furrow system. In this respect, the uplands are different from lowland areas where the second mode of scheme management prevails.

Institutions, Norms and Symbolism

The *organisation of irrigation* follows a relatively standardised model within this upland area. A slightly different organisation was reported for some of the larger derelict schemes. The leadership lies with a furrow elder or headman. This headman is said to own the furrow, and he is seldom elected. As described in the section on the construction of the furrows above, the furrows had originators or founders. The latter make the first cut in the bank of the stream and performed the necessary sacrifices to *Ruwa* (God), and thus established furrow rights (Informant M). These initial rights have been retained within the lineages of the founders, but the furrow is also said to belong to those who cultivate below it. The furrow chairman acquires his position through inheritance. The family usually decides succession on the conclusion of the funeral (*matanga*) of a deceased chairman. He should perhaps be seen as the representative of his lineage, as he is usually appointed by its members.

The position is nevertheless often inherited down a direct line of descent, since elder or junior sons can be appointed. If libations to appease the ancestors are found to be necessary in order to make the water flow steadily and smoothly, the furrow chairman is the one to perform it. This follows from his relationship to the founder. Furrow matters, in common with some other agricultural matters, have undergone a process of secularisation and disenchantment, but this remains an ultimate source for the (moral) authority of the furrow chairman, and strengthens his management rights. The aspect of mysticism surrounding irrigation is related to the ancestral cult, which has been transformed and subordinated rather than replaced by near universal adoption of Christianity. Libations for the furrow are not frequent, but few would declare that they intend to discontinue them: '*My father was the last to sacrifice. I am responsible to do it, but I have not done it yet. ... I must perform a sacrifice at the intake because the ancestors were able to get water from that intake. It is thanking God for giving us blessings*' (Informant H, furrow chairman).

The furrow chairman is the only person who has the power to call users to work on the furrow. This is adhered to even in cases when the chairman is too old to work and not even a furrow user. The organisation of maintenance is thus a core task of this office. Work on the furrow is not performed on the weekday of self-help activities, when the village organises road maintenance work among other things. Any household using the water is under obligation to send one man when requested to do so by the chairman. The chairman, who is also an expert on this technology, monitors the condition of the furrow and makes the decisions when work is deemed necessary, and also has authority over scheme layout. He has a limited control over water utilisation: '*The furrow chairman makes the decisions on cleaning. ... I make my own decisions about irrigation*' (Informant N).

Indirectly, the furrow chairmen control access to the scheme, since they call out the work parties, and also have some power to prohibit the digging of new sub-branches, and to seal leaks. These decisions can be understood as technical, but they do have an effect on access to water. An accomplished furrow chairman earns respect and invitations to beer. Formal tribute or gifts (*mashiro ha mfongo*) to the furrow elder or the group for entrance were not reported as current in the upland study area, though this had been the practice in other areas in the past. The chairman's own use of the water cannot easily be challenged, however. Some schemes have a second designated role in the furrow organisation, the overseer of the canal intake.

The people with rights to land in the command area of a furrow constitute an unnamed social group, the *mfongo* water community. This water community is an informal organisation. There is no clearly defined membership, but membership status can be attributed either to households or to male household heads, which have an equal status as participants in the scheme. Attendance in communal work strengthens the claim to irrigation water. The water community is a territorially defined unit, as clans other than that of the founding lineage for the scheme are usually represented in the command area. Command areas do not correspond closely to patrilineal clusters nor to the political boundaries of hamlet and village. The degree of correspondence varies. For one of the furrows in Kisu, most users have a shared clan allegiance different from that of the canal elder and the founder. This situation is due to immigration facilitated by a redistribution of land classified as underutilised by the Prince during British rule, and probably to land purchase in the past. Political processes beyond the hamlet thus led to a situation where most landholders in the command area of a furrow share clan name, even if their geographical and genealogical origin was actually quite distant. The creation of this new clan cluster could be interpreted as an expression of an ideology that clansmen should form compact neighbourhoods. The chance that the owner of a plot in Kisu hamlet shares clan with the canal elder of that area is 0.24. A comparison of the network of furrows and the political boundaries based on sketch-mapping of the village in 1995 shows that furrows are more likely to cross village (39%) than internal hamlet (11%) boundaries. This strengthens the interpretation that the hamlet is a basic and older unit, over which the village is superimposed.

The individual schemes are often said to belong to the furrow chairman or to his family group: '*It is our furrow. Ndorano's furrow and the one below belong to my brothers and me*' (Informant O). Chairmanship includes a free use of water, invitations to beer parties and respect, but scheme tenure does not confer tenure over water. On the contrary, water is regarded as a gift from God. The apparent interpretation is that this is a description of communal tenure. That, '*It is the water from God, and it does not belong to any particular person*' (Informant P), may be taken as equivalent to '*Everybody has got the right to use the furrow. Each and everybody needs to use it, for example for washing coffee*' (Informant N). The right to furrow water is seen as a part of the general right to subsistence, and the peasant ideology seeks to protect these entitlements. Thus, the statements relate to a morality of water use. The furrow chairman (Informant G) expressed the norms thus: '*All water is a gift from God. It is a sin to pay for water. A sin against God*'. Not to demand money for water is therefore a central norm in *mfongo* irrigation management; to mix payment and water allocation is immoral and subject to sanction. If the exchange of beer or livestock (*mashiro ha mfongo*) for access to furrow use or group membership was legitimate, this could be explained through Barth's[57] concept of economic sphere, but no informant in this particular area has chosen to report any such transactions to me. According to Johnsen,[58] payment for irrigation water was understood to be subject to supernatural sanctions, and it has been a controversial issue in irrigation settlement schemes since the 1950s, when it explained some of the reluctance to take part.[59] The insistence that the withdrawal of water without a licence from the Water Officer was illegal, the installation of flumes designed to limit the maximum abstraction of the schemes with minimum consultation with the user groups and the introduction of volumetric water fees were among the central issues in a conflict which explains the limited results of an attempt to reform water management for the Pangani Basin in the 1990s.[60]

The distribution of water is guided by a set of norms for water use. At the normative level, water use is free; everybody is entitled to reasonable use of water as a part of the right to subsistence. Some base their argument on need: '*[we] depend on furrow water*' (Informant Q). The schemes have to be kept in order, and contribution to the upkeep of the scheme is associated with access to water for irrigation. According to one of the furrow chairmen, '*Most of the time, those who attend communal work will get water*' (Informant G). Contribution to the scheme is required: '*we have the right because we maintain the furrow*' (Informant R). There is an obligation to share the common water, and no single user is supposed to divert all the water at any given time. Many users emphasised the need to '*take consideration of the other users*' (Informant Q). All matters of the furrows are rather firmly fixed in the male domain, and more strictly so than other matters in the public sphere (Informant S and others). Irrigation thus has gendered connotations. A taboo or prohibition excludes women from communal work on the furrows. According to community norms, it is also unseemly for women to turn the direction of the water or to manipulate the furrow. Further, they should not go to the intake or clean canals. Rather, it is held that they should act through male representatives in these matters. Male labour migration constitutes a challenge to these ideas.

Negotiations Over Water

In conditions where water is scarce, access to water in the case study village is the result of a continuous process of negotiation, in which normative arguments are involved. These norms do not dictate the outcome, in terms of distribution of water, which also depends on other factors, such as power, standing, and the ability to alter or break the rules or to build alliances. The increasing number of female-headed households provides a test case. According to the norms of irrigation, access to water depends on contribution to the scheme, and female participation in maintenance is prohibited. Thus, according to the normative rules, it could be expected that female-headed households are unable to irrigate. The outcome of the allocation process is shown in Table 2. The results show only a weak tendency for a higher incidence of reported irrigation in male-headed households, and the distribution shows no significant relationship. Access to water may be more difficult for female-headed households, but the distribution of reported irrigation by gender of household head is less skewed than expected on the basis of the norms alone. This can be attributed to flexibility in the application of rules. Female-household heads can be excused for non-participation, and some are able to break the rules.

TABLE 2. REPORTED IRRIGATION BY GENDER OF HOUSEHOLD HEAD

	Non-irrigator	Irrigator	Total
Male head	36% (16)	64% (28)	100% (44)
Female head	50% (10)	50% (10)	100% (20)
Total	41% (26)	59% (38)	100% (64)

(Phi = -0.13, p > 0.30)

Source: Author's Compilation

Similarly, wealth (socio-economic group) is not found to have a significant effect on reported irrigation (Phi = 0.06, p > 0.11). Two other factors are found to vary systematically with the incidence of irrigation. The first factor is affinity. Membership in the furrow-founding patrilineage (Table 3) does increase the likelihood of reported irrigation (Phi = − 0.30, p > 0.95). Contribution to the scheme is another significant factor, in accordance with the norms of irrigation. Participation in scheme maintenance is almost universal among those who irrigate permanent and field crops, but somewhat less usual among those who only water kitchen gardens with vegetables (Table 4).

TABLE 3. REPORTED IRRIGATION BY MEMBERSHIP IN
FURROW-OWNING LINEAGE

	Non-irrigator	Irrigator	Total
Same clan	31% (15)	69% (33)	100% (48)
Other clan	65% (11)	35% (6)	100% (17)
Total	40% (26)	60% (39)	100% (65)

(Phi = − 0.30, p > 0.95, *Chi-square test*: p > 0.95)

Source: Author's Compilation

TABLE 4. REPORTED IRRIGATION BY ATTENDANCE IN
WORK PARTY IN THE PREVIOUS YEAR

	Non-irrigator	Irrigator	Total
Not attended	60% (12)	31% (11)	41% (23)
Attended	40% (8)	69% (25)	59% (33)
Total	100% (20)	100% (36)	100% (56)
Not valid	6	3	9

(Phi = 0.29, p > 0.95, *Chi-square test*: p > 0.9)

Source: Author's Compilation

A specific right to a water allocation does not follow automatically from the general right, and actual performance deviates from the patterns that might be inferred from the norms and rules. The distribution of water takes place through a process which is negotiated ad hoc between users, with reference to norms and reasonability. No one is barred from drawing water by bucket. At the simplest level, the prospective irrigator simply diverts a proportion of the water flowing in the main furrow above their homestead. This is often the case outside the peak irrigation seasons.

When the amount of water in the main furrow is insufficient, the prospective irrigator has to *'go to the intake'*, i.e. They have to follow the canal upwards and close or reduce any competing abstractions on the way to the intake, and also stop leaks and adjust the intake structure if necessary. If the demand for water is high, the water could be gone again by the time the irrigator returns to their farm, due to the actions of a competing user. Two or more users take turns in depriving each other of water, and what one female-household head and irrigator described as a '*game of hide and seek*' (Informant T) often results. In her experience, '*those who are near the source get more [water]*'.

According to Informant U: '*The timing [of water application] depends on the circulation between the users… There is no schedule, but I will go to turn the direction of the water. Then I get water for an hour or so, and someone turns it again. There is a lot of running [laughs and shakes his head]. I irrigate when I can get some water. If not, I try the next day. … You have to get up early in the morning*'. He takes part in the irrigation of maize and beans in September, when the water supply is insufficient and competition for water more intense. The next step for the irrigator in need of water is to give notice to the upstream users, claim the water and close their abstractions. Conflicts are handled within the neighbourhood or the hamlet: '*The balozi, the furrow chairmen and the users are responsible. People from other areas are not concerned*', said Informant I, a retired village official.

The level of conflict over water use is usually not very problematic, since water is not so scarce in relation to demand: '*Co-operation is good, because there is more water here. It is not like in Himo and Makuyuni [lowland], where there is a scarcity*' (Informant V). According to informant W, '*Getting water from the Saranga [upland] furrow is no problem in comparison with the (…) [Lowland] furrow. In Makuyuni, there are bribes for water, and people feed their cattle on their neighbour's farm. There are no problems here*'. Conditions at home are thus contrasted with '*the other place*' down slope. The cropping pattern gives peaks of water demand for irrigation in the dry periods of August and January, and irrigators complain about difficult access to water during these periods. Conflicts frequently relate to water distribution in the peak seasons, the construction of new sub-furrows, the protection of water quality and the priority of water use for crops and other purposes.

Conflicts over water use are usually resolved between the users themselves: '*no person regulates water use. But if someone blocks the furrow, people take action and they approach him*' (Informant X). A 'sanctioning party' composed of co-users makes a visit and usually gives the culprit a scolding, or seizes their property. A female-household head who lived at the head end of a scheme and who had been subject to this method reported: '*I am not allowed to divert water by sub-furrow. … I am not allowed to make one even during the dry season. My neighbours would take action. Those from further down would come and shout a lot. I tried it. … Men are not pestered when they make a sub-furrow. … I was instructed to clean a section of the furrow*' (Informant Y).

There are conflicts of interest between the need for water for domestic use, which requires some permanent flow in all branches, and irrigation, which requires sufficient flow to be directed to one point. The furrow chairmen are able to use their authority over technical issues to stop abstractions for water-demanding crops. In other cases, elders from the area and the 10-cell leaders are mobilised and a small parley organised, often including the furrow chairman. A hamlet chairman (Informant F) reported that he had been involved in closing down a new and contested sub-furrow higher up on the canal. He went to the area together with the furrow chairman (Informants E, R and an elder from that area), and issued a warning. A few participants in the surveyed schemes reported having complained to either a *balozi*, furrow chairman or hamlet chairman. The tail-ender who had been enraged by this new branch (Informant R) provided another example. When he observed that another user was polluting the water by washing his coffee in the furrow, he went to see the furrow chairman and his 10-cell leader. '*The furrow does not belong to me, but to the other people. They do not involve me in furrow work*', said informant Z. Informant Z was an elderly widow who had to discontinue her irrigation of watercress and close the new branch, and her inability to participate in scheme maintenance was used as an argument to exclude her. It is likely that her weak position in the community made it difficult for her to draw much from the common resource.

Conclusion

Irrigation by stream diversion and spring-fed canals, locally termed *mfongo,* is a long-standing tradition on the slopes of Mt. Kilimanjaro, which has the largest cluster of 'hill furrows' in eastern Africa. The analysis of irrigation organisation has been dominated by organic metaphors and the theme of social control. Questions of process in indigenous irrigation have thus been left unaddressed, with implications on the understanding of irrigation development, scheme sustainability and access to water. The analysis of tenure and access to water provides an entry point. When the micro-politics of the canal in terms of claims argued and fulfilled and the political-economy problems described by Chambers of 'who gets what where, how, when and why' in terms of water[61] have received little attention, this is also because hydraulic tenure tends to be fluid and difficult to grasp.

This article has explored the canal organisation, according to a lineage mode, the first of the two main modes identified. The analysis has proceeded along three lines. The institutional-normative level has touched upon norms, institutions, statuses, arenas, and symbolism. The analysis of interaction has drawn on qualitative data about negotiations and how users access water. The investigation of performance or outcome has relied on quantitative analysis of reported access to water for different social groups.

At an institutional level, the canal organisation relies on the status of the furrow chairman, who is given 'ownership' in a process controlled by the family of the canal founder, which, in turn confers management rights. The status of the chairman is strengthened through devices drawn from indigenous religious ideas. The 'enchantment' of the norms serves to strengthen authority and to boost the norms of proper conduct in relation to water. Conflict resolution and technical control over work or on the scheme are more central to the chairman's position and influence, however. The scheme is associated with an unnamed and informal social group, the *mfongo* water community. Group membership is not exclusive to the kin, but depends on access to land in the command area and participation in work. The link between contribution (investment) in the scheme and the right to a share of the water is crucial in the sustainability of schemes. Agreement at a normative level can be found over leadership, the access of non-kin, not mixing money with the management of the commons, on the requirement of participation in work, and on norms that exclude women. Normative arguments are involved in the interaction, where access to water is negotiated in a continuous process, but these norms do not dictate the outcome, which also depends on factors such as power, standing, and the ability to alter or break the rules or to build alliances.

Some normative principles could be tested against reports on the practice of irrigation. When wealth rank has no effect on access to water, this is in agreement with ideas of equal opportunity to achieve rights to water. Rights can be achieved through investment in the scheme through participation in the work party, which has a moderate and significant effect on water use. Kinship, however, does have an effect on irrigation, which is contrary to these norms. No significant association could be found between the gender of the household head and the ability to irrigate, despite rules of operation which should exclude them from full participation and the acquisition of claims to water. These interesting discrepancies can be better explained if access to water is viewed as an outcome of the negotiated social process illustrated by the more detailed reports on water acquisition and conflict. The analysis of the ability to irrigate for different social groups suggests that the outcome in terms of access to water is not well - predicted from an analysis of the norms and ideology of irrigation.

This is interesting, because concerns about equity related to lineage ownership and a lack of female participation are central among the problem descriptions which have been used to validate interventions in local organisation of irrigation in Kilimanjaro Region, in addition to concerns about environmental sustainability and inadequate technology. The perception that local irrigation organisation is inferior to irrigation organisation designed by outsiders, and in urgent need of reform due to issues of equitability, or because traditional authority is eroded beyond repair may have to be moderated. Important strengths in local irrigation organisation lie in flexible principles of resource mobilisation linked to scheme tenure and the established usufruct rights and obligations. Like Fleurét[62] found for Taita, irrigation management can depend as much on multi-purpose organisations (lineage, neighbourhood and hamlet) drawing on indigenous models of organisation as on a strong single-purpose organisation (the *mfongo* water community). Neighbourhood- and kinship- based organisation can work without an elaborate set of bureaucratic rules in catering for scheme maintenance and access to water for a varied set of users, thus contributing to livelihoods and food security. The *mfongo* irrigation system has already been developed to serve new purposes, and it has been spread by Chagga migrants who replicated it, for instance in Babati.

However, the argument that lessons from local knowledge in irrigation organisation can be extracted and replicated elsewhere by actors in development, is somewhat weakened by the observation that strengths in local organisation are related to factors such as the link between investment in the scheme and rights to water, and the way hydraulic institutions are embedded in other institutions and normative systems. Nevertheless, knowledge of practice in the local organisation of irrigation can be important in the negotiation of irrigation redevelopment projects as well as water management reform processes. It may also provide some alternative models, for instance in the general meeting where access is linked to the contribution to the scheme, or in the proportional principle in water allocation, reflected in the use of time schedules and sequences, which can have advantages over volumetric principles in being easier to implement and in accommodating variable flows.

Notes

1. Early iron age in the region is c.100-1000 AD. Mturi, A. A. Archaeology of Tanzania (Dar es Salaam: The Open University of Tanzania, Faculty of arts and social sciences, 1998), pp. 118.

2. The *mfongo* (Kichagga) denotes a ditch or furrow. It is related to other hill furrow systems in north-eastern Tanzania such as the *mvongo* of the Wagweno, the *sasi* of the Vuasu and the *mtalo* of Kisambaa speakers in Usambara.

3. Sonjo, Marakwet, Pokot, Konso, among other areas.

4. The Taita hills, the Pare Mountains, the Usambaras, Mt. Meru among other areas. Plains oases around Mt. Kilimanjaro are documented by Beez, J. Die Ahnen essen keinen Reis. Vom lokalen Umgang mit einem Bewässerungsprojekt am Fuße des Kilimanjaro in Tansania. (Bayreuth African Studies Working Papers no. 2.) (Bayreuth: Universität Bayreuth, 2005).

5. Adams, W. M. "Irrigation before development: Indigenous and induced change in agricultural water management in East Africa.," in African Affairs, 87 (1988) pp. 519-535.

6. Adams, W. M. "Definition and Development in African Indigenous Irrigation," in Azania, no. XXIV (1989), pp.21-27.

7. Tagseth, M. "Genealogy as a source to the establishment of water furrows. Examples from Mt. Kilimanjaro, Tanzania," in Dahlberg, A., Öberg, H., Trygger, S., Holmgren, K. & Lane, P. (Eds.) Second PLATINA workshop 17-19 October 2002. Usa River, Arusha, Tanzania. (Working Paper from the Environment and Development Studies Unit 46.) (Stockholm: The University of Stockholm and the British Institute in East Africa), 36-37. <http://130.237.186.140/Platina/content/files/Platina%20report%20Tz/report2_www.pdf> (Accessed October 28, 2009). Tagseth, M. "Oral history and the development of indigenous irrigation. Methods and examples from Kilimanjaro, Tanzania," in Norwegian Journal of Geography 62(2008a), 9 – 22. Tagseth, M. "The expansion of traditional irrigation in Kilimanjaro, Tanzania," in The International Journal of African Historical Studies, 41(2008b), pp. 461–490.

8. Tagseth, M. "The 'mfongo' irrigation systems on the slopes of Mt. Kilimanjaro, Tanzania," in T. Tvedt, Coopey, R, Jakobsson, E & Ostigaard, T. (eds.), A History of Water. Volume 1: Water Control and River Biographies. (London: I.B. Tauris, 2006), pp. 488–506.

9. UNDP "Human Development Report 2005. International Cooperation at a Crossroads. Aid, Trade and Security in an Unequal World." (New York: United Nations Development Programme, 2005), pp. 226.

10. Kikwete, J. K. "Speech by the President of the United Republic of Tanzania, his excellency Jakaya Mrisho Kikwete, on inaugurating the fourth phase parliament of the United Republic of Tanzania." Parliament buildings, Dodoma, 30 December 2005 <http://www.tanzania.go.tz/hotuba1/hotuba/051230_bunge_eng.htm> (Accessed 18 January, 2006).

11. JICA (Japanese International Cooperation Agency). The Study on the National Irrigation Master Plan in the United Republic of Tanzania. (Dar es Salaam: The Ministry of Agriculture and Food Security, 2002).

12. JICA, National Irrigation Master Plan.

13. JICA, National Irrigation Master Plan, pp. 211.

14. Hunt, R. C. and E. Hunt. "Canal Irrigation and Local Social Organization," in Current Anthropology, 17, 3 (1976), pp. 389-398. Coward, E. W. Planning technical and social change in irrigated areas. In M. M. Cernea (Ed.), "Putting people first: Sociological variables in rural development" (New York: Oxford University Press, for World Bank, 1991), pp. 46-72.

15. Ostrom, E. Governing the commons: the evolution of institutions for collective action (Cambridge, University Press, 1990). Davidsen, T. "Playing by the rules : Cooperation, rules and water use in a Chagga irrigation system, Mount Kilimanjaro, Tanzania". (M.Phil thesis. Dept. of Geography, Norwegian University of Science and Technology, 1997). Gillingham, M. E. (1999). "Gaining access to water: Formal and working rules of indigenous irrigation management on Mount Kilimanjaro, Tanzania," in Natural Resources Journal, 39, no. 3 (1997), pp. 419-441.

16. Tiffen, M., Mortimore, M. and Gichuki, F. More people, less erosion: environmental recovery in Kenya. (Chichester: J. Wiley, 1994).

17. Fairhead, J. and Leach, M. Misreading the African landscape. Society and ecology in a forest-savanna mosaic (Cambridge: Cambridge University Press, 1996). Gillson L., Sheridan, M. J. and Brockington, D. "Representing environments in flux: case studies from East Africa.," in Area, 35, No.4 (2003), pp. 371-389.

18. Mehta, L., Leach, M. , Newell, P., Scoones, I., Sivaramakrishan, K and Way, S. "Exploring understandings of institutions and uncertainty : new directions in natural resource management." IDS Discussion Paper 372. (Brighton: Institute for Development Studies, 1999).

19. Vincent, L. Hill irrigation: water and development in mountain agriculture (London: Intermediate Technology Publications on behalf of the Overseas Development Institute, 1995).

20. Fleuret, P. "The social organisation of water control in the Taita Hills, Kenya," in American ethnologist : A journal of the American Ethnological Association, 12, (1985), pp. 103-118.

21. Bruns, B. R. and Meinzen-Dick, R. Negotiating water rights (London: ITDG publishing, 2000).

22. Moore, S. F. Social Facts and Fabrications. 'Customary' Law on Kilimanjaro 1880-1980 (Cambridge: Cambridge University Press, 1986).

23. Denzin, N. K. The Research Act: A Theoretical Introduction to Sociological Methods. (Englewood Cliffs, N.J.: Prentice Hall, 1989).

24. Rebmann, J. "Journey to Jagga." In Krapf, J. L. (Ed.), Travels, Researches, and Missionary Labours During an Eighteen Years' Residence in Eastern Africa. Together with Journeys to Jagga, Usambara, Ukambani, Shoa, Abessinia, and Khartum; and a Coasting Voyage from Mombaz to Cape Delgado. (...) (London: Trübner and Co., 1860), pp. 230-265. Kersten, O. Baron Carl Claus von der Decken's Reisen in Ost-Afrika in den Jahren 1859 bis 1865. Erster band. Erzählender Theil (New edition Graz: Akademishe Druck und Verlagsanstalt, 1978 [1st. Ed Leipzig 1869]). New, C. Life, wanderings and labours in Eastern Africa. With an account of the first successful ascent of the

equatorial snow mountain, Kilima Njaro and remarks upon East African Slavery (London: Frank Cass & co. Ltd., 3rd ed. 1971 [1st ed. 1873]. Johnston, H. H. The Kilima-njaro Expedition. A Record of Scientific Exploration in Eastern Equatorial Africa. And a General Description of the Natural History, Languages, and Commerce of the Kilima-njaro District (London: Kegan Paul, Trench and Co., 1886).

25. Odner, K. "A preliminary report on an archaelogical survey on the slopes of Kilimanjaro," in Azania, No. 6, 1971, pp. 131–149.

26. Tagseth, "Oral history".

27. The estimate of the number of schemes is based on Pangani River Basin Water Office records. The average water throughput is calculated for the catchment of the Himo River.

28. Stahl, K. History of the Chagga People of Kilimanjaro (London: Mouton & Co., 1964).

29. Wimmelbücker, L. Kilimanjaro - A Regional History. Vol 1 - Production and Living Conditions: c. 1800-1920 (Münster: LIT Verlag, 2002).

30. Moore, Social facts and fabrications.

31. Cf. Rebmann, "Journey to Jagga".

32. Gutmann, B. "Das Wasserrecht. Dargestellt nach dem Bewässerungssystem der Wamotsi und Wambokomu," in B. Gutmann, Das Recht der Dschagga. Mit einem Nachwort des Herausgebers (München: Beck, 1926) pp. 413-421.

33. Op. cit.

34. Stahl, History of the Chagga.

35. Mosha, R. S. The heartbeat of indigenous Africa: A study of the Chagga educational system (New York: Garland Publishing, 2000), pp. 61.

36. Cf. Sheridan, M. J. "An irrigation intake is like a uterus: culture and agriculture in precolonial North Pare, Tanzania," in American Anthropologist, 104, No. 1 (2002), pp. 79–92.

37. Cf. Fréon, M. "L'irrigation tratitionelle á Uru Mashariki, versant sud du volcan Kilimandjaro: un témoin de l'dentité Chagga," IFRA Les cahiers d'Afrique de l'Est, No. 26(2004), pp. 49-73. Tagseth, M. "Social and Cultural dimensions of irrigation management in Kilimanjaro," in T. A. R. Clack (Ed.), Culture, history and identity: Human-Environmental Relations in the Mount Kilimanjaro Area, Tanzania. (BAR International Series 1966) (Oxford: Archaeopress, 2009), pp. 89-105.

38. Tagseth, M. "Local practices and changes in farmer managed irrigation in the Himo Catchment, Mt. Kilimanjaro," in Ngana, J. O. (Ed.), Water Resources Management. The Case of Pangani River Basin. Issues and Approaches (Dar es Salaam: Dar es Salaam University Press, 2002), pp. 48-63.

39. Millet yields ranged from 2.3 to 3.4 tons per ha under irrigation and approximately half that amount under rain-fed conditions according to Griffith, A. W. Chagga Land Tenure Report. Moshi: [the Government Printer], 1930. pp. 45. (Document in the East Africana collection of the UDSM library [EAF CORY 272])

40. Tagseth, "Local practices and changes".

41. The furrow headman is referred to as mwenyekiti or 'chairman' in Kiswahili by the local administration, but normally this position is termed meku o mfongo; furrow headman or elder in Kichagga.

42. The agnates should be offered to purchase any plots of land which have been acquired through inheritance prior to any sale to a non-agnate. The use of pata plots for arable cultivation can be borrowed or rented for cash or an agreed amount of the harvest, while kihamba land now appears to be exclusively controlled through inheritance (and marriage). Details from interviews on land tenure history indicate that some land for Kihamba groves was allocated by the mangi during British rule, and that some few plots were transferred through purchase.

43. Komalyangoe, Marangu West Ward, Moshi Rural District in Kilimanjaro Region

44. Field inspection October 1995 and lists provided by Komalyangoe Village and Marangu West Ward.

45. Measurement by J.J. Temu, Pangani Basin Water Office in March 1996.

46. See Schanz, J. Mitteilungen über die Besiedlung des Kilimandscharo durch die Dschagga und deren Geschichte. (Baessler- Archiv. Beiträge zur Volkerkunde, Beiheft 4) (Leipzig: Teubner, 1913).

47. Fosbrooke, H. A. and P. I. Marealle, "The engraved rocks of Kilimanjaro: Part1," in Man, No. 52 (1952), pp. 161-162.

48. Odner, "A preliminary report."

49. See Tagseth, "Genealogy" and Tagseth, "Oral history", fig. 5, showing a tabulation of the lineage and its furrows.

50. See Feierman, S. Peasant intellectuals. Anthropology and History in Tanzania, (London: The University of Wisconsin Press, 1990) and Sheridan, M. J. "Cooling the Land: Politics, Ecology, and Culture in the North Pare Mountains of Tanzania" (PhD thesis, Boston University, Graduate School of Arts and Sciences, 2001).

51. In trials at Lyamungo, adding 51mm of water per month increased the output of coffee by 27%, while irrigation with mulching increased output by 107%. Wallace, G. B. and M. Wallace, M. "Magonjwa ya Mtama." Tanganyika. Department of agriculture. Pamphlet no. 55 (Dar es Salaam. Government Printer, 1953), 32. [UDSM East Africana collection: EAF per s357. a35.no3 irrigation.]

52. Howard, M. T. &. Millard, A. V. Hunger and Shame: Poverty and Child Malnutrition on Mount Kilimanjaro (New York: Routledge, 1997). Moore, Social facts and fabrications.

53. A precisely known relationship to the founder of the resident patrilineage is often claimed, and the description of this kin group as a clan can be questioned on the basis of usage in anthropology. See Keesing, R. M. Cultural Anthropology. A Contemporary Perspective (Fort Worth: Holt, Rinehart and Winston Inc., 1981), pp. 227.

54. Moore, Social facts and fabrications

55. Op cit., 139.

56. Moore, S. F. &. Puritt, P. The Chagga and Meru of Tanzania (Ethnographic survey of Africa. East Central Africa, 18) (London: International African Institute, 1977), pp. 20.

57. Barth, F. "Economic spheres in Darfur," in R. Firth (Ed.), Themes in economic anthropology (London: Tacistock, 1967), pp. 149-174.

58. Johnsen, T. R. "Wachagga: Kaffepriser og jordbruksknapphet. Problemer i et øst-afrikansk småbrukersamfunn." (Thesis for the Mag. Art. degree in Social Anthropology, University of Bergen,1969).

59. Allan, W. The African Husbandman (Edinburgh: Oliver & Boyd, 1965). Johnsen, "Wachagga", Molohan, M. J. B. Annual reports of the Provincial Commissioners for the year 1959. Northern Province (Dar es Salaam: Government printer, 1960).

60. Mung'gong'o, C. "Pangani Dam Versus The People," in A. D. Usher (Ed.), Dams as Aid. A political anatomy of Nordic development thinking (London: Routledge, 1997), pp. 105.118. Usher, A. D. "Pangani power struggle. Nordic dam builders on a Tanzanian river," in A. D. Usher (Ed), Dams as Aid. A political anatomy of Nordic development thinking (London: Routledge 1997), pp. 119-132. Gillingham, "Gaining access to water".

61. Chambers, R. "Men and Water: the Organisation and Operation of Irrigation," in B.H. Farmer (Ed.) Technology and change in rice-growing areas of Tamil Nadu and Sri Lanka (Boulder, Colorado: Westview Press, 1977), pp. 340-363.

62. Fleuret, "The social organisation of water control", pp. 115.

The Importance of Flexibility:
An Analysis of the Large-Scale Tana-Delta Irrigation Project in Kenya, Implemented Under an Estate System

Delphine Lebrun, Olivier Hamerlynck, Stéphanie Duvail and Judith Nyunja

Introduction

Although history has always been driven by "change", it seems that precisely "change" has also been the challenge to development. The constant evolution of the environment is the essence of sustainability, but also its major constraint. This is equally true for the evolution of human society. Moreover, as demographic pressure has become more intense over the past decades, discordance between environmental and social dynamics has increased significantly. On the one hand, this article portrays the idea that the environment is constantly changing, and that it is emerging as the outcome of dynamic and variable ecological processes and disturbance events, in constant interaction with human use.[1] Therefore, the availability of natural resources varies over place and in time. On the other hand, communities should be seen as a set of complex, on-going social negotiations and shifting alliances, rather than as stable, homogeneous and harmonious entities.[2] They can therefore not be treated as static, rule-bound wholes, since they are composed of people who actively monitor, interpret and shape the world around them.[3] Local communities traditionally develop livelihood[4] strategies adapted to their natural environment, in order to minimise their vulnerability.

The interaction between environmental services and various social actors is shaped by formal and informal institutions,[5] which influence and determine the access to and entitlement over livelihood assets (natural, human, physical, social and financial capital). The relationships among institutions and between scale levels (micro-, meso- and macro-) determine which social actors gain access to and control over the local resources. In turn institutions influence the uses to which resources are put and the way they are managed; and, thus progressively help to shape and modify the landscape over time.[6]

This article explores the relevance of integrating existing environmental and social dynamics for development purposes. Referring to an analysis of the Tana Delta Irrigation Project (TDIP) in Kenya, it emphasizes the importance of flexibility during the planning and the implementation of development projects. The TDIP was implemented at the end of last century and had important (and predicted) negative environmental and social consequences – working through until today. The large-scale rice project had a strong top-down implementation and did not recover from an El Niño shock (1997). It is a typical example of a failure of integration that needs in-depth analysis and understanding before introducing new projects to the same area. In fact, by granting participation to local communities and granting them ownership, the chances for effective integration of traditional local knowledge and externally imposed ruling institutions in development projects can significantly improve.

This paper is based on the "Conflict and Development" Master's thesis presented by Delphine Lebrun at the University of Ghent, Belgium under the direction of Prof. Koen Vlassenroot, coordinator of the Conflict Research Group.

Context

(a) Study Area

The TDIP scheme is situated along the Tana River, in parts of Salama Location, Garsen Division of the Tana Delta District, Kenya.[7] The Tana River arises in the Aberdare Mountains and Mount Kenya and runs over more than 1000 km to the eastern part of Kenya.[8] Due to its intense meandering and shifting character between Garsen and its estuary in Kipini, the fan-shaped delta forms a wetland ecosystem that occupies a surface of about 3000 km², fringed by sand dunes towards the sea.[9] For the local people the Tana River is the main source of water for domestic use but also provides water for a wide range of wetland-associated ecosystem services, e.g. for recession agriculture, irrigation, fishing, dry-season pasture, collection of wetland plants and animals used as food, medicine and thatching materials, reed, clay and sand harvesting, bathing, swimming and cultural practices. It provides for river transport (especially for farm and fisheries produce) and is an important security barrier against bandits, which is enhanced by the presence of crocodiles and hippopotamuses. The flood-dependent ecosystem is home to a rich biodiversity[10] dispersed in the floodplains, oxbow lakes and riverine forest patches (e.g. Masha, Kirume, Mkunumbi, Mkayumbe, Kanekeneke).[11]

Around 7000 people live in Salama Location, and they have developed a strong affinity with and possess a vast knowledge of the environment[12]. Hence, they have an age-old tradition of livelihood diversification that is based on the wetland's ecosystem services (e.g. food, water, medicinal plants, fuel wood, materials for building and handcrafts).[13] The originally agriculturalist Pokomo; pastoralist Orma and Wardei; and, the hunter-gatherer Wata tribes have developed risk-reducing mechanisms conforming to the rains and the bimodal flooding regime of the Tana River. This enables them to use different resources from the same ecosystem. By doing so, they avoid the risk of becoming over-dependent on one single resource.[14] Floods are perceived as being essential for the sustenance of floodplain fertility, and therefore supports the farming system. To these tribes, the floods are also vital for the productivity of most of the natural resources on which these communities depend i.e. forestry, fishery and wildlife.[15]

The long rainy season is from April to June, and the short rains occur from October to December. The floods reach their maxima in May and November. May flows are generally higher and less variable than the November flows. The lowest river flows occur in February-March and September-October.[16] During the flood seasons, communities mainly rely on fishing and flood recession agriculture on the floodplain, while their main activity during the wet seasons is rain-fed agriculture.[17]

In their study of the Pokomo and Wardei tribes in the Lower Tana, Terer et al.[18] describe how the tribes practice shifting cultivation, grow maize, bananas, mangoes and rice in the flood plains, oxbow lakes and riverbeds. To optimally exploit the brief flood recession period, they plant rapidly maturing crops such as millet and sorghum. The cultivation of mangoes along the river remains an important source of income for the local agricultural community.

Fishery is practiced in the river and its oxbow lakes. Local people understand the reproductive cycles and population dynamics of fish and hence take this knowledge into account while fishing. The locals, therefore, adapt their fish harvesting strategies (e.g. by shifting to meat) according to their observations in order not to overexploit specific fishery or fishing grounds. Seasonal rotation in fishing grounds between the lakes and the river is adopted. Decision making on activities to be carried out at a particular time is based on discussion and consultation. The Pokomo have strong clan alliances (*milalulu*) and lineages by which elder's decisions and opinions are highly respected.[19]

It is important to link what happened within the TDIP with the broader river basin context. The construction of hydroelectric power dams between 1968 and 1988[20] in the upstream parts of the Tana substantially reduced peak flows.[21] Maingi and Marsh also noticed a decrease in run-off and meandering of the Tana, and the related deposition of sediments. This resulted in an overall reduction of the floodplain fertility. Emerton,[22] states that the decreased flooding regime also led to a disruption of the traditional patterns of transhumance. It also increased the grazing pressure and it has intensified the conflicts between pastoralists and floodplain agriculturalists over land and the use of resources on the banks of the Tana.[23]

Methodology

Three out of the six villages in the TDIP scheme were selected: Baandi, Hewani and Vumbwe[24]. The three other villages, Wema, Kulesa and Sailoni, are all inhabited by the Pokomo and are not as thoroughly discussed in this analysis. Baandi, Hewani and Vumbwe were chosen because of their contrasting characteristics contributing to diversity in perception by the communities.

Hewani is a Pokomo village of about 150 households. Pokomo are mainly agriculturalists. Like the people from Vumbwe, those from Hewani settled on the present-day location after the extreme floods of 1961. In 1969 the village started its own irrigation system: the Hewani Minor Irrigation Scheme (HMIS). HMIS and Wema Minor Irrigation Scheme (WMIS) are the only village irrigation projects that still exist within the TDIP today.

In Vumbwe there is a Pokomo community of 20 houses (312 people) and a Wata family that consists of 7 houses (24 people).[25] The Wata people used to rely principally on hunting and gathering but nowadays they mainly make a living from farming.

Baandi is the only Orma village (204 houses, 1000 people) within the TDIP. Although Orma people are traditionally pastoralists they have been increasingly combining this activity with agriculture since their settlement in the Tana Delta.

The analysis was done through participatory research in the selected villages, based on the sustainable rural livelihoods framework.[26] In the three villages, group discussions and semi-structured interviews were held in order to construct a livelihood and environmental entitlements profile. The participants were randomly selected (on the basis of availability and willingness to participate), after addressing the Headman in every village asking for representatives of various stakeholders. At the same time, resource maps and timelines were developed, and transect walks were made with local farmers.[27] Several employees and ex-employees from TARDA, working in the estate compound in Gamba or in Queensway House in Nairobi, and some ex-employees from Garsen were also interviewed.

The Analysis

Generally, the information gathered in Baandi, Hewani and Vumbwe can be divided into four consecutive time periods separated by events or landscape changes linked to significant phases in the history of the TDIP. Based on these findings, an analytical framework (cfr. Table 1) was elaborated, describing the *'natural capital'* of the TDIP (i.e. the natural environment and the infrastructure), the *'livelihood strategies'* of the local communities, their *'participation'* in and their *'perception'* of the TDIP. These factors were analysed over the periods: *before* and *during* the TDIP, *after* the El Niño (1997) and in relation to the *future* projects.

The "before TDIP" refers to the years (20th century) before the start of the implementation of the project in 1988[28] and thus to the period prior to the construction of the main embankment. This embankment was built around the Blocks A to G (1763 ha prepared for irrigation) of Polder 1 North and fenced off an area of about 2400 hectares. As will be described later on, the new infrastructures (the embankment, main canal and secondary network, the TARDA estate compound with a rice mill complex[29], a hospital – that has been turned into a school -, the 1st, 2nd and 3rd class housing for TDIP employees, tarmac roads, a mobile phone network, etc.) induced several changes in the study area "during the TDIP". However, by the end of 1997 East Africa was hit by El Niño-related floods, which created chaos and destroyed major parts of the recently completed TDIP infrastructure, especially in its the southernmost part, while Blocks A, B, C and D remained relatively undamaged. The consequences of this event are clarified in the "after El Niño" period. As a consequence, 1998 was marked by important food shortages (the rice of block G was lost entirely), displacement and numerous cholera and malaria victims. Although the Japanese initially proposed to fund the reconstruction (July 1998), they left the project soon afterwards. In the last phase of the analysis, the "new projects and the future perspectives" are put forward. To conclude with, alternatives are discussed and recommendations are made.

In this analysis, a first major question is whether the negligence of the dynamics of resources and means of development has an influence on the sustainability of livelihoods.[30] A second and related question is whether changes in environmental entitlements induce changes in sustainable livelihoods. Answers to these questions are quite apparent as they have been argued elsewhere, but they are once more illustrated through this participatory research in the failed large-scale TDIP in Kenya.

The relevance of this case study lies in its effort to provide insight into the perception of the advantages and disadvantages of the irrigation project by collecting facts and views from different stakeholders. By discussing the participation and perception of local communities in and about the TDIP, this paper also draws attention to certain institutional changes that determine the access and control over the resources, to examine how the locals were consulted and considered during the irrigation project. Allowing local people to participate provides them with the opportunity to express their needs and interests.

This information is important to compare the project's achievements and failures. In this manner lessons learned from past mistakes can, and should, improve the implementation of future projects. By drawing attention to past failures in the Tana Delta, it aims at informing and warning stakeholders and decision makers of the traps and problems that can hamper successful implementation of future projects. Moreover, the recent drought conditions in Sub-Saharan Africa and Kenya, climate change and demographic pressure exacerbate the growing importance of sustainable water management projects. Last August 25, 2009, the Kenyan Director of the World Food Programme drew attention on the fact that *"Life has never been easy for the poor in Kenya, but right now conditions are more desperate than they have been for a decade".[31]* Questions as 'what are the factors that lead to sustainability and success of projects?' and 'when can a project be considered successful?' are therefore highly relevant.

Four Steps in the Evolution of the TDIP

Before describing the evolution of the TDIP, it is interesting to provide some background information about the project financing and its organisation. At the request of the Kenyan Government, the TDIP was financed by the Japanese Government through the Japanese Overseas Economic Cooperation Fund (OECF). Japanese bilateral assistance consisted of both financial and technical aid[32]. An agreement between the OECF and the Tana and Athi River Development Authority (TARDA) was signed in March 1987.[33] TARDA was and remains the

Kenyan parastatal institution in charge of the implementation of the project from a local management office in Gamba (within the TDIP estate compound) but the Board of Managing Directors is based in Nairobi. The managing office in Gamba operates with a current and a savings account. The income from production must be deposited in the savings account, and nobody in Gamba has access rights to this account. All financial decisions require the approval from Nairobi, while the managing office in Gamba is not consulted for all decisions taken in Nairobi.[34]

The TDIP has always operated with an "estate system" (with exception of one seasonal "tenant system" experiment), under which all agricultural development activities -including agricultural infrastructure, production, quality control, marketing and sales, administration, operations and maintenance- are managed and administered in a unified manner. This means that the management and execution of all processes such as production, harvesting and sales, etc., was conducted by the TDIP Office.[35] In other words, ploughing, harrowing, sowing, the provision and distribution of seeds, water, fertilizer and chemicals was carried out by the TDIP itself, while the community members guaranteed weeding, bird and wildlife scaring etc. through casual labour contracts.

As mentioned before, the implementation of the TDIP induced changes in various aspects. These changes are synthesized in the analytical framework below and described further in subsequent paragraphs.

Table 1:	Analytical framework of the TDIP			
	Before TDIP	During TDIP Implementation of Polder 1 North (Block A to G) start in 1988	After El NIÑO '97-'98 Japanese stop funding in 1998	Future - New projects
Natural capital	Frequent inundations & consequent migrations. Diversions of Tana River (e.g. '80). Oxbow lakes, riverine forests, lots of wildlife. Regional: Construction of hydroelectric power dams in upstream parts of Tana between 1968 and 1988. Roads, blocked during flooding seasons. Hewani, Wema, Vumbwe and Sailoni at actual location.	↓ water level in Tana river36, ↓ peak flows, meandering & related deposition of sediments, with an overall ↓ of the floodplain fertility as consequence37. Construction of main TDIP embankment starts in 1989. Then: main canal & network, headworks with rubber dam at Sailoni. Diversion of Tana at Mnazini. Drying out of lakes. Improved roads & accessibility, mobile phone network & infrastructure (hospital, housing, offices, rice mill, etc.) Baandi at actual location (inside the embankment). Degradation of riverine forest, ↓ number & diversity of wild animals, dying of perennial crops along the river banks due to higher water table.	Changes in Tana river course & low water level. Degradation of riverine forests, deforestation, drying out of lakes, ↓ availability of natural resources, ↑ threats by wildlife. 80% of infrastructure completed in 1997. 1999: TARDA starts temporary construction: Main TDIP road & embankment partially restored (earthen), head canal, sluices & secondary canals rehabilitated from Sailoni to Kulesa/ Wema ('99). 2005: Restoration Malindi-Lamu tarmac road by Chinese. November 2007: rupture of inflatable rubber dam. Since 2007 all surrounding lakes have completely dried out. ↓ fertility of village shambas. 2008-2009: Sugarcane experiment for TDISP at Kulesa (Block A).	If no action: continued deterioration of TDIP infrastructure due to vegetation grow & siltation. More research needed. Planned rehabilitation of rubber dam by September 2009 (Update December 2009: still didn't happen). TDIP Youth Project. TDISP by Mumias Sugar Ltd. Jatropha project by Bedford Biofuels? Horticulture project by Government of Qatar? Projected Grand Falls dam?

Table 1:	Analytical framework of the TDIP			
	Before TDIP	**During TDIP Implementation of Polder 1 North (Block A to G) start in 1988**	**After El NIÑO '97-'98 Japanese stop funding in 1998**	**Future - New projects**
Livelihood strategies	Great diversity of risk coping mechanisms. Combine: flood - fed agriculture on floodplains, rainfed agriculture & Small scale irrigated agriculture: LTVIP38; Hewani, Wema and Kulesa Minor Irrigation Schemes. Livestock keeping. Fishing (Tana & Oxbow lakes), Waterlelies & firewood collection, charcoal, hunting, beekeeping, trade.	↓ diversity of livelihood strategies. ↓ fishing opportunities. ↓ amount of water lilies and species of medicinal herbs. ↑ access to food (certainly rice). ↑ opening to regional market. ↑ job opportunities. ↑ cash income & flow of money.	↑ distances for water collection, fishing or access to food. ↑ reliance on firewood winning & charcoal burning. Last TDIP harvest in 2003. Irregular payments by TARDA, getting worse with the years. Sometimes payments in rice instead of money. ↓ cash income & money flowing. ↓ TDIP employments39, ↓ job opportunities. Since rupture of rubber: complete loss of irrigation. Forced to rely on rainfed agriculture, while climate change disrupts general meteorological conditions.	

Table 1:	Analytical framework of the TDIP			
	Before TDIP	**During TDIP Implementation of Polder 1 North (Block A to G) start in 1988**	**After El NIÑO '97-'98 Japanese stop funding in 1998**	**Future - New projects**
Ancestral trustland or government land: customary entitlements. Laws restricting hunting since 1970s. Good access to grazing areas. Feasibility study ('83 & '87), EIA ('85), Research by Nippon Koei (started in 1986). No direct consultation of local population about TDIP.	TARDA claims itself owner of TDIP land without effective title deed (only an Allotment Letter from the Commissioner of Land, valid for 3 months). People found out about the TDIP through construction of infrastructure. Villages threatened by destruction during construction of irrigation scheme. TDIP estate system: no entitlements to local population. Loss of land ownership by local communities, loss of access to grazing land (warranted by confinement and temporary imprisonment). New institutions about access to water. Complete restriction of village expansions due to TDIP scheme boundaries. 1995: "tenant system experiment" Start Civil Case no. 660 in 1995 against TARDA at Mombasa High Court.	Communities have no right over fallowing TDIP plots. Constant postponing of Court Case for reasons known by TARDA. Case still pending today. When people report problems, no efficient measures are undertaken. Claiming for payments are not effectively granted by TARDA. Pressure from TARDA to keep decision making in own camp.	Feasibility study & EIA for TDISP. EIA approved by NEMA. TARDA is looking for funding. Multiple EIA comments on Mumias TDISP, a.o. by EAWLS40 & Nature Kenya. Case running against Mumias Ltd. at Malindi High Court. Claims for tenant system instead of estate system: it would give people more ownership over the scheme.	

Table 1:		Analytical framework of the TDIP		
	Before TDIP	**During TDIP Implementation of Polder 1 North (Block A to G) start in 1988**	**After El NIÑO '97-'98 Japanese stop funding in 1998**	**Future - New projects**
Perception	Little certainty due to seasonal floods and wetlands. Access to great diversity of natural resources. Important traditional knowledge about ecosystems & environmental goods & services. Nomadic life for many Orma & Wardei populations, with low number of school - going children.	↑ living standard surroundings TDIP. Protection from floods by embankment. Improved transport & communication facilities. ↑ financial independency for women, ↑ ability to make own household decisions. ↑ sedentarisation of pastoralists related to creation of job opportunities & growing desire for education. ↑ time for children to go to school. Gradual loss of traditional knowledge. ↑ conflicts between agriculturalists & pastoralists. Many unmet promises by TARDA, e.g. no grazing corridor within the scheme, educational and health facilities not uplifted to projected level, no building of bridge between the TDIP & Garsen, etc. People are "squatters on their own land". ↑ illness & diseases, especially malaria due to paddy rice.	↓ living standard surroundings TDIP. Impossibility to return to original diversity of livelihood strategies. Conditions worse than before project. High unemployment rates, certainly among youth. Rehabilitation only for 30%, deteriorated irrigation & protection from floods. People would not turn to poaching as such if they still had fishing. Fading of traditional knowledge among youth.	EIA Mumias Ltd. is not representative enough. Consultation of uneducated people from villages outside the TDIP (Garsen, Ngao, Witu and Kipini). TARDA neither consults villages before, nor during or after taking action. People want & need more livelihood opportunities & better access to food & water. Necessity of job creation & better marketing opportunities. Lack of educational facilities (general & environmental) & awareness programs. "Food crop versus business". People are more familiar with the process of rice than with sugar cane. Rice is more labour consuming & would create more jobs.

Source: Author's Compilation

Before the TDIP

Before the TDIP, the course of the Tana River was locally hardly controlled by human intervention and life was highly regulated by the river dynamics. From the different group discussions it appeared that migrations of entire villages and clans were often a necessity due to extreme floods. Before 1900 the three largest villages in the actual TDIP area were referred to as Gadeni, Rubenmwewe and Ababidu. In 1958 people from Ababidu were forced to leave because inexplicable deaths were threatening the villagers. Together with families from Chunoni and Chikosi, two other nearby villages, they created Kulesa. Heavy floods in 1961 also caused many people to migrate, which resulted in the creation of Hewani (and Shika Lako),[41] Wema (and Langu Moyoni),[42] Vumbwe and Sailoni. The most significant flooding in the twentieth century took place in 1914, 1961, 1980 and 1997. They all led to migrations. The recent floods in 2007 caused food insecurity and a high number of cholera victims. They also forced the people from Baandi to live around Gamba for several months.

Communities from the actual TDIP area used to rely on different lakes in the surrounding wetlands: e.g. lakes Chamadho, Mulanja, Mkuju or Mukuyuni (Pokomo vs. Wata language), Roka, Jange, Mtsenkwa or Masenkwa (idem), Mtwapa, Ngunu.[43] Water lilies, collected in those wetland lakes, were a rich source of food, as well the roots as the flowers and the grains were used for different types of recipes.[44]

Local people have also taken advantage of traditional small-scale irrigation. An example is the Lower Tana Village Irrigation Project (LTVIP) which was funded by the Dutch Government through the Kenyan Ministry of Agriculture in the 1980s. This project was aimed at implementing the minor irrigation schemes of Oda, Ngao, Hewani, Wema and Mnazini.

In 1983 a draft EIA report to TARDA already predicted numerous problems but the TARDA refused to recognize these findings. The report drew attention to the risk of the disruption of Orma pastoralists' use of the rangelands and the interference with their ability to water their livestock; the toxicity of the mercury-based biocide "panogeen" (sprayed aerially) and the threat to the fragile, disappearing riverine forests and two endangered monkey species.[45]

During the Project Implementation

The construction of the TDIP infrastructure introduced a permanent and significant transformation in the project area. Water supply for the irrigation of the farm was regulated through a network of supply and drainage canals around and in between of the different TDIP Blocks. Water from the Tana River was diverted through a head canal -North of the farm- and subsequently tapped by a major sluice gate at Sailoni, composed of an inflatable rubber dam and a system of sluice gates to regulate the flow into the canals of the irrigation scheme. Since the construction of the embankment and the extraction of water for irrigation, the communities noticed a decline of the water level in the surrounding lakes.

This caused a visible degradation of riverine forests. During flood peaks, on the contrary, the embankment causes the water table outside the TDIP to rise, which destroys perennial crops like bananas, mangos, sugarcane, paw paw, etc. along the river banks. These effects can also be related to the decline in number and diversity of wildlife in the TDIP area (elephants, ostriches, antelopes, giraffes, buffalos, hippos, crocodiles, waterbucks, etc.).[46]

However, according to quite a lot of locals, the TDIP had improved the living standard of local people and surrounding villages during the productive years. They were better protected from the seasonal floods, which brought uncertainty. The livelihood outcomes had improved: more food security, better transport and communication facilities, financial independence for women and more time for children to go to school. The increased sedentarization of pastoralists could be attributed to the creation of jobs and the growing desire for education. Nevertheless, Pokomo and Orma tribes claim that the gradual loss of traditional knowlegde among their communities is due to these changes to their life pattern.

Both the transportation and communication facilities improved considerably thanks to the project, and this also advanced the access to information and the marketing opportunities in the area. "*Before the TDIP, transport used to be restricted six months per year, as the floodplains were inundated during April-May-June and November-December-January*".[47] The TDIP had opened up the area to regional markets. It was the best rice in the region, and it was sold to cities as distant as e.g. Malindi and Mombasa. During the production years (from 1993 to 2003) the access to food – certainly rice- had improved. Some women expressed that "*as opposed to now, there was no shortage of food in the area, there was business to do and money was flowing. When the message was spread that combine harvesters were working on the TDIP, people would come from different places – even from Hola and surroundings- to collect the remaining rice on the plots. This harvest could go up to 50 kg/person. TARDA did not complain about that. They only objected when people went to the fields that had not been harvested yet. After the production, some women also collected the little rice pieces leftover within rice husks. This would also provide us with entire bags*".[48] Though this was locally perceived as a positive impact of the TDIP, collecting rice grains from a harvested field can hardly be considered as an indicator of major improvement of livelihoods

The project created jobs, making cash income to play a more important role in livelihoods. However, local people were only employed on a casual basis for jobs such as watchmen, bird scarers and water regulators. For many women it was the first time they enjoyed some financial independence. This gave them the opportunity to make their own household decisions and significantly increased their self-esteem. In Hewani and Vumbwe only one person was employed permanently while nobody enjoyed this statute in Baandi until 1998.[49]

For the TDIP project, TARDA implemented a top-down approach and managed it under an estate system from their headquarters in Nairobi. Even though preliminary studies had been accomplished, many people claim that they only found out about the project once the constructions had started. The local population had not been consulted directly. During the construction works, certain parts of the villages were threatened to be destroyed as a result of which "*people had to take up their sticks and knives for an armed blockade*".[50] Roads and embankments were designed in such a way that they would cut through the settlements, as if no-one was living there. This indicates once more the lack of knowledge and consultation of the project area and its communities.

The land in question was County Council 'trust' land. Local communities have no actual title deed, even though they have used the land for decades. Informal or customary property rights as such are legitimized by social norms and codes of behaviour. These are legitimate in the eyes of those local resource claimants who regard government-reserved land as ancestral farmland, but they are not recognized by the state in modern legal terms.[51] Within the context of the TDIP, TARDA obtained an Allotment Letter (reference number 106798 of January 1995[52]) from the Commissioner of Lands - which is valid for 3 months - but has no effective title deed.[53] Nevertheless, the parastatal ran the project as if they were the legal owners of the trust land. Communities lost their ownership over the land and the access to the resources in it: "*We became squatters on our own land*".[54] The estate system denies all entitlements to the local people. In 1995, the communities took TARDA to the Mombasa High Court (Civil Case No. 660) to challenge the land allotment to TARDA. The case is still pending, and no statement has been made because of absences in court. "*Other reasons only known by TARDA*" have also caused the hearings to be postponed every time. Today, most of the plaintiffs, essentially village elders, have already passed away.

In 1995, TARDA agreed to adopt a "tenant system experiment" for one season. During that season the farmers did the agricultural work – irrigation, weeding, scaring of wildlife and birds - while the other services, such as ploughing, harrowing, broadcasting and seeds, water, fertilizer and chemical application, were provided by the TDIP. A great part of the labour was also mechanized. At the end of the season, the harvest went up to 3.5 metric tonnes/ha, whereas from 1993 to 1995 the TDIP productions resulted in 2.5 metric tonnes/ha. Once harvested, the rice was sold to TARDA, the price being negotiated. Even though a proposal to prolong this idea was taken to Nairobi, no agreement was obtained.

The Years After the El Niño, Up to Now

The El Niño: Crucial Turning Point
Obviously, the El Niño caused changes in the river course and the damage to the infrastructure obstructed the irrigation possibilities. Although the floods boosted the fertility of the area, the continued decline of the water level in the Tana, induced riverine forest degradation. Moreover, according to Snoussi et al.[55] the current trends in the dammed catchment suggest that the physical adjustments to the water fluxes are incomplete and that the downstream, deltaic, and coastal sea environments will continue to change as a result of damming. Little is known

about the probable long-term consequences of the current damming and water abstractions.

The strong dependence on natural capital (water, soil, hydrological cycle, etc.) of livelihood strategies in the Tana Delta has been described earlier. It has become clear that the TDIP infrastructure and environmental shifts have also had an influence on livelihood strategies of various communities.

Even though significant amounts of money were made from the rice production between 1993 and 1997, the project did not recover from the El Niño shock and people could not return to their original diversity in livelihood strategies. Environmental changes have caused diminished fishing opportunities. Before, people fished in their own neighbourhoods, while today they are forced to explore more distant areas (e.g. roughly 25 km to Lake Moa). Similarly, water lilies are no longer found as easily and the number of species of medicinal plants have reduced. Since the project focused on rice production, the traditional diversity of risk-spreading mechanisms was narrowed down. This also forces local communities to rely on other types of livelihoods: *"Many people would not turn to poaching if they still had fishing".[56]* The changes in livelihood opportunities intensified the deforestation and increased the number of people falling back on poaching. Access to food has become problematic.

Due to a deteriorating relationship between TARDA and the Japanese sponsors, payments for casual labourers started to become irregular, and this grew worse over the years. It was not uncommon for payments to be made in rice instead of money. People even described the present-day living conditions as *"worse than before the project".[57]* The rising unemployment rates and strongly decreased flow of money in the area lowered the standard of living in the TDIP villages. Some young, jobless men claimed *"TARDA was not able to restart the project after the El Niño. There was no money available to pay the employees, thus, salaries were promised to be provided after the harvesting and selling of the rice. However, no agreements were fulfilled, salaries were lowered and when it turned out that these could not be paid, rice was handed out instead… until the whole thing collapsed. Until today many people claim payment for their work".*

People interviewed state that one of the most important reasons for the conflicts between TARDA and the communities is that TARDA has not kept its promises. There are plenty of examples. A grazing corridor was promised to be incorporated in the scheme, but this never happened. Pastoralists lost the access to their traditional grazing areas (TDIP plots now) and would have to go around the TDIP in order to reach the surrounding grazing areas. This is, however, difficult, knowing that it takes seven hours to bring livestock from Baandi to Vumbwe (from 6 am to 1 pm). Pastoralists who took their livestock through TDIP plots, which increased the risk of conflicts, were confined and temporarily imprisoned. Neither Educational nor health facilities reached the promised level. A promised bridge over the Tana that would connect the TDIP to Garsen still has not been built. A lack of written contracts between TARDA and the communities that outline the nature and details of the relationship surrounding the TDIP, is one of the causes of these impossibility to challenge these unmet promises.[58]

Unfortunately, in November 2007 a mistake during the inflation of the dam caused a fissure in the inflatable rubber, which had been used intensely by the local farmers to irrigate village *shambas*. The irrigation of the TDIP area (including the Minor Irrigation Schemes) became almost impossible after this incident. This resulted in lack of water flowing through the irrigation channels, reduced soil fertility and drying up of nearby lakes.

The diminished availability of natural resources causes more conflicts between people and wildlife (e.g. buffalos, hyenas, elephants, lions, among others) in the vicinity of the project area. For example, people complain about wild animals that use their village boreholes to drink. The Mumias Ltd. sugarcane experiment on Block A at the height of Kulesa (for the Tana Delta Integrated Sugarcane Project (TDISP)), also brought about more hinder from buffalos. Since the rupture of the rubber dam, people are forced to rely on rain fed agriculture causing food insecurity the current declining rainfall conditions. The loss of traditional livelihoods have led to intensified charcoal burning and firewood collection by the local people.

New Projects and Future Perspectives

At present, the TDIP infrastructure is deteriorating further, due to vegetation growth and sediment deposition in the TDIP canals. Transect walks through the Blocks C, D, E and F showed expansion of Acacia and Prosopis in the abandoned fields. Developments have come to a standstill, though a rehabilitation of the rubber dam is planned by September 2009 and casual employees are not being paid correctly due to a lack of funds. Neither the project, nor the parastatal are currently beneficial for the Kenyan economy. In fact, in April 2009, watchmen were striking in order to obtain their pay. TARDA is looking for funds and several private companies have already shown their interest in the TDIP area for the production of sugar cane (e.g. MAT International in 2004).[59]

In 2007 Mumias Sugar Ltd. engaged in the TDIP, while the Environmental Impact Assessment (EIA) of the Tana Delta Integrated Sugar Project (TDISP) was approved by the National Environmental Management Authority (NEMA). Nonetheless, many comments on the Mumias TDISP and its EIA (e.g. by Nature Kenya and the East African Wildlife Society[60]) have been raised and discord spread among the communities. The comment of the EAWLS on the EIA[61] was based on an exhaustive study of the EIA itself and different public meetings held in Kulesa, Wema, Hewani, Baandi, Gamba, Danisa and Sailoni[62]. According to EAWLS, significant negative impacts with serious ecological, social and economic ramifications at local, national and international levels have been identified but no serious implementable mitigation measures have been provided. The loss of the prime floodplain grazing land of the delta is one of the main impacts of the TDISP. In case of drought the Tana Delta is the safety net for more or less all the cows between the Somali border and Malindi. Not only will the grazing land occupied by the TDISP be lost, but also all the areas excluded from flooding by the embankment and by the reduction of flooding frequency linked to the abstraction of a substantial proportion of the flow.[63]

Similar to the TDIP, the local people were not sufficiently consulted about the project. TDIP villagers claim that uneducated people from villages outside the scheme (Garsen, Ngao, Witu and Kipini) were consulted for the TDISP EIA. Moreover, Mumias Sugar Ltd. was summoned at the Malindi High Court by members of the communities and the conservation organisations in Nairobi, to dispute the land ownership issue. Likewise, local people continue to face broken promises and vague arrangements. *"The land within the TDIP went from the communities to the parastatal, and now it will be handed over to the private sector. The local people do not understand that they will lose their land. Unfortunately, due to the lack of funds, it is difficult to undertake awareness campaigns properly".*[64] The local communities strongly need empowerment.

According to Scoones this includes the expansion of assets and capabilities of poor people to participate in, negotiate with, influence, control, and hold accountable institutions that affect their lives. Increased livelihood opportunities and better access to food and water are needed. There is an important need for jobs (certainly among the younger generations) and better marketing possibilities. Educational facilities are not sufficiently available and people ask for environmental education and awareness programs. Moreover, many people state that they are more familiar with rice than with sugarcane ("food crop versus business"). Those who prefer sugarcane base it on their negative experience with the TDIP for rice.

Alternatives

The views and facts outlined in this article clearly add up to the idea that negligence of the dynamic of resources and means of development have an influence on the sustainability of livelihoods. They also demonstrate how changes in environmental entitlements induce changes in sustainable livelihoods. The changes perceived along the TDIP are synthesized in Table 2.

Lessons learnt from the failed large-scale Tana Delta Irrigation Project in Kenya should be considered for future decision making. The case study illustrates the importance of integrating ruling environmental and social dynamics in development purposes, in order to contribute to the sustainable improvement of the living standard of communities while mitigating the environmental impacts.

If management decisions in the Tana Delta continue to be taken top-down, without participation of local communities; people in the area will continue to face poor development perspectives. Instead of implementing business-driven projects based on agricultural intensification (of one crop), an integrated livelihood-based approach should be applied, supporting livelihood diversification and traditional knowledge. The livelihoods of the locals have to be analyzed in depth, with the aim of enforcing people's capabilities and empowering the most vulnerable groups.[65] Therefore, local farmers and pastoralists have to be consulted on their needs and interests. People should not only be granted participation in project planning, but also in project implementation. It is necessary to give them responsibilities and ownership (environmental entitlements). This enables them to organize themselves and to take up their development opportunities. They should be empowered to make their own livelihood choices. Obviously, development opportunities have to be created, for example through access rights to sustainable managed resources or through development of small-scale irrigation.[66] That is also why the TDIP would work out better under a tenant system.

TABLE 2: AN OVERVIEW OF THE CHANGES

Means to achieve	Before TDIP	During TDIP	After El Niño '97-'98	Future
Natural & Physical Capital	Lifestyle adapted to dynamic of environment	TDIP to allocate the Tana river & control floods	Destroyed infrastructure, Changed hydrology	Continued attempts to further allocate Tana & control floods
Livelihood Strategies	Livelihood diversification & Migrations	Agricultural intensification & permanent settlement. Casual labour	Loss of irrigation & difficulty to turn back to livelihood diversification.	Restoration of diversification or further intensification? Casual labour?
Institutions	Informal property rights, trustland.	Loss of ownership & environmental entitlements Formalization of institutions. Changed power relations, more stakeholders	Formal rules above informal customary institutions	Further loss of ownership and access?
Vulnerability Of Local Communities	High	Reduced	Higher & increasing	Currently no perspectives of improving
Perception Of Population	Hard life due to wetland dynamic	Ameliorated livelihood outcomes but suppression of rights	Worse than before project	Poor perspectives, Empowerment necessary.

Source: Author's Compilation

Development projects have to remain within a flexible and transparent framework. Flexibility should be present as well in the planning, as in the management and in the implementation stages. The static, large-scale TDIP infrastructure is not suitable for such a type of dynamic wetland ecosystem.

Previous large-scale irrigation projects (e.g. the failed Bura and Hola Irrigation Schemes upstream of Garsen) also proved to be inefficient in the Tana River Valley.[67] Such inflexible irrigation schemes do not correspond to an integrated river basin management system as necessary in the Tana Delta. More attention should be paid to the successful small-scale Minor Irrigation Schemes in the area. They can be combined properly with livestock grazing and more assets can be made available for other types of livelihoods. Nevertheless, if large-scale projects are preferred, they should be submitted to continuous monitoring and evaluation, responding to the environmental and social dynamics.

Conclusion

Japanese development aid to Africa was initially motivated by the need to ensure access to raw materials. During the implementation period of the TDIP Japanese aid was rapidly expanding, partly due to Japan's need for diplomatic support in its bid for a permanent position on the UN Security Council.[68] At the same time other donors had mostly abandoned the playing field in Kenya because of frustration.[69] In addition, the high capital outlay of the TDIP infrastructure was highly compatible with the tied nature of donor aid and greatly benefited Japanese consultancy and engineering firms. At the same time Kenya was rather desperate to keep donor money flowing and be seen as being on a development course in spite of the departure of its traditional donors.

Both partners therefore probably had a less than critical attitude about the TDIP and chose to ignore what was by that time already a well-established truth from countless experiences all around Africa, namely that the exclusion of the main hypothetical beneficiaries of the intervention, i.e. the local communities, from its planning, implementation, running and ownership would lead to its failure quite independently from its accidental destruction by the El Niño floods. The lack of any credible strategy for the rehabilitation of the scheme points to the low economic and social credibility of this approach to development. The same issues were pointed out for a Japanese funded irrigation scheme in Malawi.[70]

A more holistic and participatory approach, developing an optimal set of flood-dependent ecosystem services that can ensure the improvement of the various local livelihoods, including fisheries, forestry, livestock keeping and gathering in addition to irrigated and recession agriculture would be likely be more economically, environmentally and socially sustainable. Rehabilitating the hydraulic infrastructure and reconnecting the forest patches as proposed by the Critical Ecosystem Partnership Fund[71] would be a first step. Using parts of the water to resupply the lakes and restore the fisheries would be another and would have immediate beneficial effects on wildlife by reducing the monetary incentives for poaching and other detrimental activities. Ironically the final result of such an approach would probably closely resemble the much - vaunted Japanese Satoyama landscape[72] that has supported a functional combination of substantial forest cover, grazing areas and rice fields while maintaining high biodiversity over many centuries on communal lands.

Notes

1. Leach, M., Mearns, R. and Scoones I. (1999). Environmental Entitlements: Dynamics and institutions in community-based natural resource management. World Development Vol. 27, No. 2, pp. 225-247.

2. Allison, E.H. & Badjeck, M.-C. (2004). Livelihoods, local knowledge and the integration of economic development and conservation concerns in the Lower Tana River Basin. Hydrobiologia 527: 19-23.

3. Leach et al., 1999.

4. A livelihood comprises the capabilities, assets (including both material and social resources) and activities required for a means of living (Scoones, 1998).

5. Scoones, I. (1998). Sustainable Rural Livelihoods: A framework for analysis. IDS Working Paper 72. According to Scoones institutions are the regularised practices (or patterns of behaviour) that are structured by the rules and norms of society which have persistent and widespread use. They are thus often fluid and ambiguous. They are usually subject to multiple interpretations by different actors.

6. Leach et al., 1999.

7. The administrative hierarchy in Kenya is constituted as follows: Province, District, Division, Location, Sub-Location.

8. The Tana River basin holds about 20 % of the national population, a major portion of the agricultural potential and the highest hydroelectric power generation potential in the country (Maingi and Marsh, 2002).

9. Maingi JK, & Marsh SF (2002). Quantifying hydrologic impacts following dam construction along the Tana River, Kenya. *Journal of Arid Environments* 50: 53-79.

10. Noteworthy is the presence of species incorporated in 'The IUCN Red List of Threatened Species' e.g.: two species of endangered primates, the Tana River Red Colobus *Procolubus rufomitratus rufomitratus* and the Tana River Mangabey *Cercocebus galeritus* with a 'decreasing population trend' according to Jong Y.A de & Butynski T.M. 2009. Primate Biogeography, Diversity, Taxonomy and Conservation of the Coastal Forests of Kenya. Unpublished Report to the Critical Ecosystem Partnership Fund, 97 pp. + annexes. The Tana River poplar *Populus ilicifolia* is classified as 'Vulnerable' (WCMC, 1998).

11. These forest patches were explicitly named by the local communities because they noticed important degradations over the last years. They have been represented schematically during the participatory mapping, but have not been georeferenced.

12. Personal communication, Chief of Salama Location, May 4, 2009

13. Hatfield, R., Cunneyworth, P. and Luke, Q. (2005). "Rehabilitation of the Tana Delta Irrigation Project, Kenya – an Environmental Assessment." [Unpublished report, Critical Ecosystem Partnership Fund].

14. Terer et al., 2004.

15. Duvail S., Hamerlynck O. (2007). The Rufiji River flood: plague or blessing ? *International Journal of Biometeorology*. Special issue on climate perception. Vol. 52, n° 1, pp. 33-42.

16. Maingi and Marsh (2002) and Snoussi M, Kitheka J, Shaghude Y, Kane A, Arthurson R, Le Tissier M, & Virji H (2007). Downstream and coastal impacts of damming and water abstraction in Africa. *Environmental Management* 39: 587-600.

17. Terer et al., 2004.

18. Terer et al., 2004.

19. Terer et al., 2004.

20. Seven Forks hydropower plant': Kindaruma dam (completion year 1968), Kamburu dam (1974), Gitaru dam (1978), Masinga dam (1981) and Kiambere dam (1988). ISRIC, World Soil Information (2006). Green and blue water services in Tana river basin, Kenya: Exploring options using an integrated modelling framework [Draft]…

21. Maingi JK & Marsh SF, 2002.

22. Emerton, L. (2003). Tana River, Kenya: Integrating downstream values into hydropower planning. Case studies in wetland valuation # 6. Document produced under the project "Integrating Wetland Economic Values into River Basin Management" from DFID and IUCN.

23. Kagwanja PM 2003. Globalizing Ethnicity, Localizing Citizenship: Globalization, Identity Politics and Violencein Kenya's Tana River Region. *Africa Development*, Vol. 28 : 112–152.

24. Baandi' derives its name from the TDIP protective "band" where the Orma people settled in 1989, to be protected from the floods. 'Hewa' means breeze in Pokomo language, Hewani is thus "place with a breeze". 'Vumbwe' is named after a little pool that the villagers found when they arrived there in 1961.

25. In the last few years they sometimes share their land with a clan of Wardei pastoralists –who were there at the time of the field visit-. However, they are not included in the analysis as they were never involved in the TDIP. Generally speaking, Wardei arrived in the area after the El Niño.

26. DFID, 2001. Sustainable Livelihoods guidance sheet 4. URL: http://www.dfid.gov.uk and http://www.nssd.net.

27. Ibid.

28. The feasibility study was elaborated by Haskoning Royal Dutch Consulting Engineers and Architects in November 1982 and updated in 1987, while the EIA study was finalized in July 1985 (with draft reports in December 1982 and October 1983) (Hirji, R. and Ortolano, L., 1991 Strategies for managing uncertainties imposed by environmental impact assessment. Analysis of a Kenyan River Development Authority. *Environmental Impact Assessment Review* 11, pp. 203-230.).

29. After its completion, the rice mill was tested during one week: 700 mT/ hour could be milled. Apart from that week, the machine was never used, and the Japanese became suspicious about the way TDIP money was spent (Pers. comm. by an ex-employee of TARDA, 21st of April 2009).

30. According to Scoones (1998) a livelihood is sustainable when it can cope with and recover from stresses and shocks, maintain or enhance its capabilities and assets, while not undermining the natural resource base.

31. UN News Centre, Kenya (August 25, 2009). UN agency sounds alarm on dire food situation. URL: http://www.un.org/apps/news/story.asp?NewsID=31842.

32. Hirji, R. &Ortolano, L. (1991). The financial support of the TDIP was coordinated by the OECF, an affiliate of the Japanese Foreign Ministry. They financed the TDIP through a loan from the Japanese Bank for International Cooperation (JBIC). The technical assistance to the TDIP (e.g. monitoring of works, approval for disbursement of funds) was provided through the Japanese International Cooperation Agency (JICA). JICA introduced Nippon Koei Co., Ltd. Consulting Engineers as the concern that was responsible for upgrading the Haskoning Royal Dutch feasibility study in 1987 -done by Haskoning itself in 1982- and preparing the detailed project design.

33. Rowntree K. (1990). Political and administrative constraints on integrated river basin development: an evaluation of the TARDA, Kenya. *Applied Geography*, Vol. 10, pp. 21-41. The Tana River Development Authority (TRDA) was created in 1974 after the grand ideal of the Tennessee Valley Authority, with the aim of developing the water resources of Kenya. It would be an advisory body to the Kenyan government, in charge of the planning, coordination and monitoring of projects in the Tana River Basin. However, the provisions for implementing projects increased with time. In 1981 TRDA expanded geographically and broadened its mandate to the "Tana and Athi River Development Authority (TARDA)". For project funding TARDA depends mostly on multilateral and bilateral development

agencies, while the Kenyan Government generally funds the local cost of TARDA's projects (Hirji and Ortolano, 1991).

34. Pers. comm. ex-employee of Nippon Koei, May 5, 2009.

35. JBIC (2008). Aid effectiveness to infrastructure: A comparative study of East Asia and Sub-Saharan Africa. Case studies of Sub-Saharan Africa. *JBICI Research Paper* No. 36-3.

36. The optimal water level at Masinga dam is 1050-1052 m but nowadays it is at 1032m: decrease of 20m (Pers. comm. from an Engineer from TARDA, 13th of May 2009).

37. Maingi and Marsh, 2002.

38. The Lower Tana Village Irrigation Project was funded by the Dutch Government through the Kenyan Ministry of Agriculture. The aim was to implement the minor irrigation schemes of Oda, Ngao, Hewani, Wema and Mnazini. The Mnazini Minor Irrigation Scheme is not in use anymore because of a natural diversion of the Tana, redirecting the water utilized for irrigation (Pers. comm. by M.K. Diwayu, 5th of May 2009).

39. TARDA reduced the number of personnel at the TDIP office from 233 in December 1997 to 68 (JBIC, 2008).

40. This East African Wild Life Society comment on the Mumias Ltd. EIA was formulated based on different meetings held in Kulesa, Wema, Hewani, Baandi, Gamba, Danisa and Sailoni (Personal communication on 6th of May 2009).

41. People coming from abandoned Gadeni (Pers. comm. from a farmer, May 8, 2009).

42. People coming from abandoned Rubenmwewe. "Wema" means 'nice place' (Pers. comm. from a farmer, May 8, 2009).

43. These lakes were explicitly named by the local communities because they dried out over the last years. They have been represented schematically during the participatory mapping, but have not been georefered.

44. Pers. comm. by farmers, April 24, 2009.

45. Hirji, R. and Ortolano, L., 1991.

46. Pers. comm.: interviews, transect walks and participatory mapping Vumbwe.

47. Pers. comm. ex-employee of TARDA, May 5, 2009.

48. Pers. comm. ex- casual employees of the TDIP, May 3, 2009.

49. After the El Niño one person from Baandi was permanently employed.

50. Pers. comm. by farmers, April 24, 2009.

51. Leach et al., 1999.

52. EAWLS (2008). Comments on the Tana Delta Integrated Sugar Project Environmental Impact Assessment Study Report.

53. In order to acquire a title deed after an Allotment Letter, a payment has to be made (Pers. comm. ex-employee of TARDA, April 21, 2009)

54. Pers. comm. from farmers, April 22, 2009.

55. Snoussi et al., 2007.

56. Pers. comm. from farmers, May 7, 2009.

57. Pers. comm. locals from Garsen, April 30, May 3 and 4, 2009

58. Hatfield et al., 2005.

59. MAT International decided not to go through with the sugar cane project because local communities were not supportive. There is a Court Case pending between MAT International and TARDA for breach of agreement (concerning the title deed) (Pers. comm. ex-employee from TARDA, May 5, 2009).

60. EAWLS, 2008.

61. EAWLS points out several specific comments on the TDISP EIA : 1. No analysis of alternatives of project sites; 2. Land tenure/ownership issue; 3. Water balance problem; 4. Insufficient community involvement; 5. Problematic relocation of people; 6. No development of a Master Plan (integrated land use planning) for the Tana Delta; 7. Habitat fragmentation and loss of biodiversity; 8. Seawater influx/saline water intrusion; 9. Eutrophication and chemical use; 10. Displacement of livestock; and 11. Introduction of exotic species, especially oil palm (EAWLS, 2008).

62. Pers. comm. farmer, May 6, 2009

63. Nature Kenya (2007). Comments on the Environmental Impact Assessment study report for the proposed Tana Integrated Sugar Project in Tana River and Lamu Districts, Coast Province, Kenya (Land Allocation Reference No. 106796 of 17.1.1995).

64. Pers. comm. farmer on May 6, 2009.

65. For example through education, training, technical assistance, etc.

66. Nature Kenya, 2007.

67. Ledec, G. (1987). Effects of Kenya's Bura irrigation settlement project on biological diversity and other conservation concerns. *Conservation Biology* 1: 247-258.

68. Sato, M. 2005. Japanese aid diplomacy in Africa : an historical analysis. *Ritsumeikan Annual Review of International Studies*, 2005. Vol.4, pp. 67-85

69. Svenson, J. 2003. Why conditional aid does not work and what can be done about it? Journal of Development Economics 70: 381-402.

70. Veldwish, GJ, Bolding, A. & Wester, P 2009. Sand in the engine: the travails of an irrigated rice scheme in Bwanje Valley, Malawi. Journal of Development Studies 45: 197-226.

71. Hatfield et al., 2005.

72. Kobori H. & Primack RB 2003. Participatory conservation approaches for Satoyama, the traditional forest and agricultural landscape of Japan. Ambio 32: 307-311.